Interstellar Communication

Scientific Perspectives

Interstellar Communication

Scientific Perspectives

Cyril Ponnamperuma

A. G. W. Cameron

Houghton Mifflin Company / Boston

Atlanta / Dallas / Geneva, Illinois
Hopewell, New Jersey / Palo Alto / London

523.13082
P797

Printed in the United States of America.

Library of Congress Catalog Card Number: 73-11945

ISBN: 0-395-17809-6

Preface

To our ancestors a few hundred miles was almost infinity. Today we are in a world whose frontiers are receding from us almost before we reach for them. What was fantasy a generation ago, we now know as possible.

With the beginning of the space age, man's thoughts have turned to the question of intelligent life on other planets, around other suns, in other galaxies. Over a decade ago a group of scientists interested in this question assembled at the National Radio Astronomy Observatory in Green Bank, West Virginia and discussed the problem from many points of view. Their thoughts and observations were made available to the interested reader in the collection of papers edited by one of us (A. G. W. C.) under the title *Interstellar Communication*.

Although since that time the question of intelligent life beyond the earth and the possibility of communicating with extraterrestrial civilization was discussed in several papers (see Bibliography), no combined effort had been made to bring together the latest thinking on the subject. With this objective in mind a series of lectures was organized at the NASA Ames Research Center in the summer of 1970.

The subject matter of the lectures was designed to cover every aspect of the problem in a stepwise manner—from the question of the origin of the universe to that of the origin of life, the origin of intelligence and the means of communication. Since the main purpose of the series was to outline methods of possible interstellar communication, the bulk of the lectures was addressed to this problem. The others were intended to provide the background necessary and give some assurance that the quest was a scientifically plausible one.

Since there was such widespread interest in the lectures given at the Ames Research Center, it was considered opportune to provide a readable version of the presentations in a form accessible to the student or the general reader. The lectures were, therefore, reedited by the individual authors to suit the format of this book. In the course of reediting some new material has been included. However, on the whole, we have tried as much as possible to preserve the vivacity and excitement which can only be conveyed by the spoken word.

We are grateful to each of the authors for generously providing of their time and effort to make this venture a success. We owe a special debt of thanks to Dr. Linda Caren who at very short notice put together for us the Bibliography and spent much time in gathering the illustrations for this book. We also thank Dr. Robert Jastrow of the NASA Goddard

Space Flight Center and Columbia University for his careful detailed reading of the manuscript. Finally, we acknowledge with much appreciation the help given by members of the Houghton Mifflin Company, who first saw the value of making these lectures available to the student and the inquiring public.

It is our hope that the ideas contained within these pages, though at times controversial, will help the reader in a quest of truly cosmic significance.

Cyril Ponnamperuma
A. G. W. Cameron

Contents

Interstellar Communication
Scientific Perspectives

1

An Introduction to the Problem of Interstellar Communication[1]

Carl Sagan   *Cornell University*

Man's Place in the Universe

This chapter examines some current thinking on the subject of extraterrestrial intelligence and on the problems to be encountered in the attempt to detect and communicate with extraterrestrial civilizations. We begin with a short historical perspective of man's evolving ideas about his place in the universe.

Primitive man imagined himself as being rather insignificant in the universe. He believed he was living on a flat earth beneath an inverted bowl called the sky. Attached to this inverted bowl—or at least moving within its vicinity—were the sun and moon. These bodies rose and set, the latter changing phases of illumination in some regular but poorly understood way. Above the sun and moon were the stars, which were called *fixed* because they retained a constant geometrical relation to each other as they collectively rose and set. However, there were a few "stars" that were not fixed but wandered with respect to the other stars. These were called *planets*, from the Greek word for wanderers. This was a nice and tidy world. (Note that the word "world" is still used today to mean universe. Only a few hundred years ago men recognized only one world in the universe.) The objects in the sky were not viewed as *places*—they were another category of *thing*.

As science advanced, views changed, and it became apparent that the stars were distant suns that were dim not because they were intrinsically faint, but because they were so far away. It became clear that the planets were not just points of light, but rather were other worlds more or less like the earth. Speculations arose that some of the stars might have planets orbiting them; indeed, that some of the planets in this or other planetary

[1]Some of the work discussed in this chapter was supported by NASA Grant NGR 33-010-101.

1

systems might have life similar to that on the earth. In the 18th century speculation of this sort was very popular, with serious writers even postulating a different kind of organism peculiarly suited to its environment for each planet in the solar system.

Since then our knowledge of the universe has grown enormously, and it has become apparent that the universe we live in is not friendly and earth-like. Indeed, it is quite different from the earth and indifferent to man.

The Earth As Viewed from Space.   Perhaps the best emotional perspective on man's significance in the universe can be obtained from an examination of the earth as a planet. From even the nearest planets, Venus and Mars, the earth would appear merely as a bright point of light in the sky, essentially indistinguishable from the other planets. On closer observation, or with the aid of a modest telescope, one would see something like the view shown in Fig. 1.1. In such pictures, which are

Figure 1.1   The full earth; Apollo 17 photograph. Africa, Madagascar, and the Arabian Peninsula can be seen.

superior to the best ground-based photographs of Mars, the clearest features are clouds. The existence of continents and oceans is not at all obvious, nor would it be without some further and rather careful study. Only with vastly better instrumentation—in particular, a space vehicle sent close to the earth—would better resolution be possible. Figures 1.2 through 1.9 illustrate the appearance of the earth from such a vantage point.

Figure 1.2 is a typical photograph of the earth as viewed from space; the planet is 50 percent cloud covered. Note that the features usually considered the most obvious—continents, cities, forests—are not apparent even in this closer view. (Figures 1.3 through 1.9 show cloud-free regions.)

Surely there must be life in the region shown in Fig. 1.3. There must be inhabited areas, outposts of civilization, all kinds of plants and animals. However, with the resolution used in these photographs, such indications of life and civilization are not apparent. We should remember that this resolution is comparable to all but a fraction of the best resolution photography obtained by space-probes of any planet to date.

Figure 1.2 Cloud patterns off Baja California. The Gemini 5 spacecraft is in left foreground.

Figure 1.3 Typical Gemini/Apollo photograph of populated regions of the earth at about 100-meter resolution. No sign of life is evident.

A quite remarkable feature—a giant "corkscrew" on the earth's surface—is apparent in Fig. 1.4. It is probably a dry stream bed. This feature is certainly quite striking and does not appear to *belong* there. It is not a defect in the photograph—it is something on the earth's surface. It is exactly this type of phenomenon that would suddenly bring us up short if we were doing photographic reconnaissance of the earth and had not previously suspected that there might be something like life on the planet. Indeed, we might well say, "My goodness, look at that! That's pretty regular and geometrical. How did it get there?"

Southern California in the vicinity of the Salton Sea is shown in Fig. 1.5. This sea, which is the result of an error in the rerouting of the Colorado River at the beginning of this century, would be a sign of intelligent life to visitors from another world if only they knew how to interpret it. In a manner of speaking, it is an artifact of intelligent life. But perhaps the message would not be so difficult to read. Just imagine a visitor in earth orbit photographing this area once a month and noting nothing like a sea. Then one month a sea is visible. He would say, "That's very interesting. Where did that come from?" There is additional evidence of intelligent activity in this picture, both to the right (north) and left (south) of the Salton Sea. The larger area on the left is the Imperial Valley. The image is just on the verge of being resolved; if the resolution were just a bit better, a remarkably geometrized pattern would be apparent. But in Fig. 1.5 the rectangular blocks of the irrigated farmland remain tantalizingly on the verge of resolution.

Figure 1.6 shows a small city. There is clearly visible a set of straight lines (the major streets and highways) and a tendency for the feature to be set up in a rectangular array—a definite checkerboard pattern. But again, the pattern is not startlingly evident. All the photographs shown

Figure 1.4 Apollo 9 photo of the mouth of the Colorado River. The bright corkscrew is apparently a tidal drainage channel outlined by salt flats.

Figure 1.5   Gemini 5 photo of the Salton Sea (right) and the Imperial Valley (left) of southern California.

in Figs. 1.2 through 1.6 are of roughly the same resolution—some tens to hundreds of meters on the surface of the earth.

Figure 1.7 shows us quite an unambiguous geometrization. There is a remarkable frequency of squares and rectangles—something very hard to explain geologically. The pattern is, in fact, due to the tendency of human beings toward territoriality and Euclidian geometry. Even if you were not aware of these human peculiarities, you would probably be struck by the unusual regularity in the surface features. Such organization would be hard to explain except as an artifact of intelligent life. The same phenomenon appears again in the photograph in Fig. 1.8.

Figure 1.9 is a view of the Nile Delta. The dark area in the photograph indicates vegetation. Each of the small white spots is a clearing in the vegetation where people live; the straight lines connecting spot to spot are roads carved out by long usage.

Notice that with the resolution of the preceding photographs, the only detectable signs of life are the large-scale artifacts of intelligence. In all these photographs only intelligent life on the earth was indicated; conversely, before the evolution of hominids, few of the features mentioned would be visible.

There are two kinds of lesson to be learned from a study of this sort. One is that at 1-km resolution the earth has had its overall appearance altered only slightly by life and intelligence. That is, it would be difficult to find unambiguous evidence at 1 km for life such as that presently inhabiting the earth—at least if the searching is done photographically. (There are other methods of looking that will be described below.) The second point is that the visible details at 100-m resolution are primarily the products of a technical civilization that has been on the earth only a short time compared to the age of the planet. Had our imaginary visitors dropped by the earth anytime earlier than about 30,000 years ago, they would *not* have seen most of the features described above. There comes a time, if one can make any extrapolation from the history of

Figure 1.6 Dallas and Fort Worth, Texas at 100-meter resolution; Apollo 6 photo.

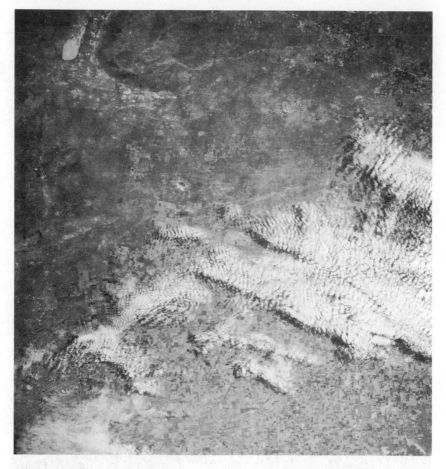

Figure 1.7   An example of the resolution of the earth into a check-erboard pattern when viewed at better than 100-meter resolution.

the earth, when life develops to the point where it suddenly makes itself visible to close-up photography, and to other observations.

The Sun As Viewed from Space.   Let us now develop a similar perspective for our parent star, the sun. The sun is situated in a lens-shaped galaxy, some 100,000 light-years in diameter, called the Milky Way. The solar system lies far out towards the edge of the disk, some 30,000 light-years away from the center of the galaxy. Ours is a most unprepossessing locale. It is hardly the first place someone would visit if he came here from another galaxy. Our part of the Milky Way contains a fair amount of gas and dust, and, as a result, there are a few young stars in the neighbor-hood. Some of the stars in the night sky are quite young in the sense of having been born from the interstellar gas and dust only in the last tens or hundreds of millions of years, perhaps a few even more recently than that. Another characteristic of our neighborhood is the relative sparcity of stars; the average distance between them is about 6 light-years. There are other places—for example, the center of the galaxy, where the average distance between stars is considerably less. This fact too may be of importance when considering the prospects for contacting other civilizations.

Men are accustomed to thinking of the sun as exceptionally bright, the ultimate source of heat and light. The sun as seen from even the nearest stars is an insignificant pinpoint of light. From the more distant stars a human eye could not see the sun at all—it is intrinsically far too faint. The insignificance of the sun and the solar system leads one, in a natural way, to think that inhabited planetary systems might be the rule rather than the exception. But this is a guess and not a scientific argument.

Figure 1.8 Another example of the resolution of the earth into a checkerboard pattern when viewed at better than 100-meter resolution.

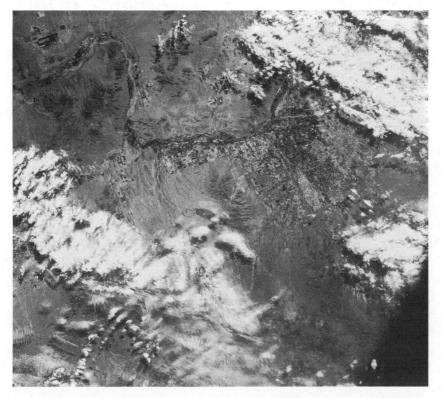

Figure 1.9   Gemini 5 photo of the Nile delta.

The Search for Extraterrestrial Intelligence

There are several questions involved in the usual formulation of the problem of searching for extraterrestrial intelligence: How many civilizations are there likely to be in the galaxy? How far away are they? What form will they have? How might we communicate with them? The search strategy one adopts is also important, and we should note that there are some ways in which ordinary astronomical research may be relevant to the question of the search for extraterrestrial intelligence.

In the last ten years, the subject has gained a growing respectability, and it is even possible to do some calculations. Our ideas about extraterrestrial intelligence are no longer in the Sunday supplement or science-fiction categories. Perhaps the primary factor in this modernization has been the remarkable development of the science of radio astronomy. The very great difficulties encountered in detecting life on earth by photography from the distance of Mars or even with a space vehicle in earth orbit have already been noted. But there is a quite trivial way in which human activity can be detected from Mars without even leaving that planet—trivial, that is, given a rudimentary familiarity with radio astronomy. A radio telescope of about 2-ft aperture (1/500 the size of the largest dish available) tuned to some meter band frequency would, when pointed at the earth (particularly North America), receive a blast of radio emission stronger than that from the sun. Continued scrutiny would probably indicate that this emission was not purely random noise but had some mild sense to it. The terrestrial emission so readily detectable from Mars has three sources: (1) domestic television transmission, (2) FM and the high-frequency end of the AM broadcast band that occasionally leaks through the ionosphere, and (3) the radar defense networks of the United States and the Soviet Union. It is a sobering thought that the only signs of intelligent life on the earth which are detectable over interplanetary and interstellar distances are housewives' daytime serials, the rock-and-roll end of the AM broadcast band, and the semiparanoid defense networks of the United States and the Soviet Union. That is it. Those are the only signs of life on earth. Perhaps this is the answer to the oft-asked question of why other, more advanced, civilizations in the galaxy have not come calling. Now we know.

The radio spectrum of the earth as monitored from Mars is shown schematically in Fig. 1.10. Since the earth is at a temperature of roughly 300°K, we would expect it to have an emission spectrum following the blackbody curve, except for some absorption by the atmosphere. Further, the curve *should* be centered at about 10μ in the infrared and fall off inversely as the square of the wavelength toward very long wavelengths. What we *find* is a truly remarkable peak in the meter band. It is not

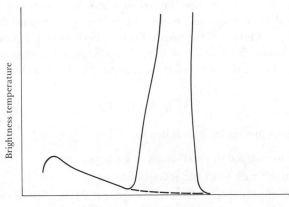

Figure 1.10 Schematic representation of the emission spectrum of the earth.

quite true to say that the earth emits more energy as radio waves than as all other types of thermal emission; yet the amount of radio emission is extremely striking. Indeed, it is so large that if one observed only at meter-band frequencies and assumed the emission there to be of thermal origin, the deduced (blackbody) temperature of the earth would be about 40 million °K. This is an example of extreme disequilibrium. That is, the earth's radio radiation is not characteristic of thermodynamic equilibrium. The strong disequilibrium component is due to the activities of the modicum of intelligence residing on the planet. Such departures from thermodynamic equilibrium radiation are, in principal, a means of searching for intelligence elsewhere. Perhaps a good place to start would be to search for planetary transmissions that are not necessarily intended for external consumption, similar to our own domestic TV and radio. Radio astronomical techniques are now sufficiently advanced to give us reasonable hope of success in such a venture, provided that another civilization is close enough.

Radio astronomical techniques have advanced to the point where communication over enormous distances is now possible. Indeed, using the largest radio telescope available (the 1,000-ft radio telescope of Cornell University at Arecibo, Puerto Rico) and the most sensitive detection and transmission devices, we could communicate with a similar telescope anywhere in the galaxy. This means that even with our present level of technology we have the ability to communicate with 200 million stars. Two questions are immediately posed by these facts: (1) Is it probable that at least one other technology coexists with us in this volume? (2) If one or more such technologies are present within communicating distance, how can we go about discovering their location and establishing communication?

To answer the first question let us first ask how many civilizations might reasonably be expected to coexist in the galaxy and follow this with a simple calculation of the mean distance between civilizations resulting from a random distribution of the calculated number over the volume of the galaxy. The usual form of the equation for the number of civilizations is:

$$\mathcal{N} = R_* f_p n_e f_l f_i f_c L$$

where the terms are defined as follows:

R_* = mean rate of star formation over galactic history

f_p = fraction of stars with planetary systems

n_e = number of planets per planetary system with conditions ecologically suitable for the origin and evolution of life

f_l = fraction of suitable planets on which life originates and evolves to more complex forms

f_i = fraction of life-bearing planets with intelligence possessed of manipulative capabilities

f_c = fraction of planets with intelligence that develops a technological phase during which there is the capability for and interest in interstellar communication

L = mean lifetime of a technological civilization

Let us now examine each term in more detail and estimate its value.

Rate of Star Formation, R_*. The more frequently stars are formed, the more likely it is that the galaxy will contain inhabited planets. Since there are about 10^{11} stars in the galaxy, and the age of the galaxy is about 10^{10} yr, R_* must be approximately 10 stars/yr throughout the volume of the galaxy. This value is probably a lower limit to the actual value when averaged over the lifetime of the galaxy. We expect the rate to have been considerably greater in the past and to be smaller at the present time.

Fraction of Stars with Planetary Systems, f_p. A primary implication of the definition of this term is that the analysis is being restricted to life of our general sort, namely, life that resides on planets. Other, nonplanetary life forms are imaginable—for example, life on cool stars or in the interstellar medium. However, the suggestion of life in or on stars with surface temperatures of a few hundred degrees Kelvin runs into insuperable thermodynamic difficulties. The light arriving on the earth comes from a source at 6000°K and drives a biology at a temperature of about 300°K. Viewed as a heat pump, the Carnot efficiency of the system is about 98 percent. In the cool-star situation, a hypothetical biology would have about the same temperature as the source from which it derived its energy —an exceedingly difficult proposition for a heat pump.

With regard to the numbers of planetary systems, we should first note our own situation. It provides not just one example but four, since the satellite systems of Jupiter, Saturn, and Uranus demonstrate dynamic characteristics rather remarkably like those of the planetary system. This suggests that the process of formation of smaller bodies of 10^{-3} to 10^{-5} the mass of the primary body is a common event. The patterns of spacing between secondary bodies are also similar in the planetary and satellite systems. The various theories of the origin of the solar system seem to argue for this as well—at least the noncatastrophic theories do. The currently fashionable theories hold that planets form at roughly the same time as their parent stars and from the same general material. Not too much weight should be attached to theories without experimental support. However, in this case such support may possibly be available.

Beyond the Alpha Centauri triple-star system, the next nearest star is Barnard's star, about 6 light-years distant. The search for planets of this star cannot be accomplished merely by looking for an additional tiny point of light nearby because the amount of light reflected by a body so close to its star is indistinguishable from the light of the star itself. This is true even for very large planets quite some distance from their parent stars such as our own Jupiter. A second method of searching would be to look for spectroscopic evidence of a blue or red shift in the light reflected by the planet, but again, this is impossible because the spectrum of the parent star swamps any spectrum from the planet. A third method is available and has been tried with apparent success on Barnard's star. This method can be used if a star is no more than a few hundred times the mass of its largest planet, if the two are widely separated in comparison to their sizes, and if the orbital plane is not in the line of sight. Under these conditions the common orbital motion of star plus planet will appear as an oscillation in the position of the star, even though the secondary object—the large planet—is not itself visible. In addition, stars show some motion across our line of sight as they travel along their orbits about the galactic center. The combination of these two motions—across the line of sight and about the center of mass— produces, over a long period of observation, a wavy or sine-wave motion with respect to the background of "fixed" or more distant stars. The American astronomer Peter van de Kamp has photographed Barnard's star (the most rapidly moving star as seen from the solar system) for several decades and has reported just such slight perturbations in the linear motion of Barnard's star across our line of sight. The two planets van de Kamp has deduced orbiting Barnard's star have roughly the mass of Jupiter, but Barnard's star has only about one-tenth the mass of the sun.

It would be extremely difficult for this technique to be applied to planets of the size of the earth—they would have 10^{-4} or 10^{-5} the mass of the star they would be orbiting. If we were on a planet orbiting Barnard's star and observing the sun, such wobbles here would not be detectable using exactly the same procedure. This is because in the Barnard system the star/planet mass ratio is about ten times larger than the ratio of mass between the sun and Jupiter. It becomes more difficult to detect such wobbles with distance and with slower motion across the line of sight. For these reasons we do not have very good hopes of applying this technique readily to very many other star systems. However, we do know that almost half of the dozen or so nearest stars have dark companions of approximately one to ten times the mass of Jupiter. This result, which is on firmer observational grounds than the results on Barnard's star, is exceedingly important.

Given these data it seems reasonable to suppose that the fraction of stars with planets is near unity. The actual number is more likely one-half, since nearly half of all stars are binary systems in which we might reasonably suppose conditions would be rather unfavorable for the formation if not the evolution of planets defined as suitable for the origin and evolution of life. Perhaps the slight optimism implied by adopting unity for f_p will compensate for the conservatism adopted for R_*.

Number of Planets Suitable for Life, n_e. In the solar system there is obviously at least one such planet. In the minds of many scientists, Mars and Jupiter also possess suitable environments for the origin if not the advanced evolution of life. At the very least the conditions on those worlds do not automatically exclude the presence of life of some form. However, n_e concerns planets in other systems where the gross conditions may not be at all similar to those in the solar system. For example, a very bright star with all the planets clustered nearby implies a number of very hot planets possibly unsuitable for life. A very dim star with most of its planets very distant results in the opposite situation. However, because of the greenhouse effect, there will still be congenial temperatures on very distant planets. If the orbital spacings of the planets in other systems are similar in absolute value to the spacing of the planets in the solar system, then there should be several planets suitable for life around most stars. This is true even for Barnard's star, which has a luminosity about 1/2,500 that of the sun. A choice of unity for n_e would again seem conservative; perhaps two, three, or even four would be nearer to what one would expect from our own situation and on the assumption that a law of geometrical spacing similar to that which obtains in the solar system is also applicable elsewhere.

Fraction of Planets with Life, f_l. If we begin with any mixture of the gases constituting a primitive planetary atmosphere and then supply this mixture with energy of almost any form, the result will be the formation of the basic molecules of which life on Earth is constituted—the amino acids and the nucleotide bases and sugars. It is rather remarkable that the laws of physics and chemistry are such that the fundamental biological organic molecules are produced in high yields at rapid rates under the most general primitive conditions. To be sure, this production of amino acids and nucleotides is not the same as the origin of life. But the production of these building blocks is a major step in that direction. Indeed, given that we can produce quantities of such materials in the laboratory in a matter of weeks or even hours, it seems possible that given a billion years of such conditions such molecules will self-organize into some very simple, self-replicating polynucleotides, weakly coding sequences of catalytic polypeptides. This latter sequence of experiments is yet to be performed, but the available evidence strongly suggests that the origin of life should occur given the initial conditions and a billion years of evolutionary time. The origin of life on suitable planets seems built into the chemistry of the universe. For the probability f_l, I choose a value of unity.

Fraction of Life-bearing Planets with Intelligence, f_i. Given the origin and evolution of life on a planet, what is the likelihood that intelligence will arise? Having no experimental evidence to support a general theory as in the case of chemical evolution, we are reduced to our own example. On the other hand, the definition of intelligence cannot be too restrictive, since dolphins and other *Cetacea* are certainly intelligent by any reasonable standard. However, they do not possess manipulative organs and so could not, even if so inclined, construct a technical civilization. Many evolutionary biologists are struck by the selective advantage of intelligence. That is, the intelligent organism is much better able to survive and adapt than his competitors, even given rather serious relative deficiencies in other areas such as sight, hearing, mobility, and so forth. But other evolutionary biologists are equally struck by the tortuous path leading to the origin of man. The number of fortuitous accidents which had to occur at the right time for man to develop the way he has is truly astronomical. The response to this objection is that we are not referring to human beings when we discuss extraterrestrial intelligence; we are, instead, discussing a class of much more general intelligent beings and recognize that details of structure and evolutionary timescale will be determined by the environment in which the intelligent species develops. The five-fingers-on-each-hand chauvinism that many of the more conservative thinkers adopt is usually invoked to show that life is unlikely to exist elsewhere in the universe, except possibly on a planet precisely identical to the earth. Such thinking is hardly Copernican in spirit. It is possible that the coin could be turned to argue that current thinking about extraterrestrial intelligence is guilty of nucleic-acid chauvinism or even carbon chauvinism (which are perfectly valid charges in the light of the preceding discussion).

Intelligence has developed several times on the earth and in a time equal to about one-third to one-half of the main-sequence lifetime of the sun. From lack of any information beyond the assumption of our own mediocrity, I believe we should adopt a value near unity for f_i; this amounts to stating that intelligence is an inevitable consequence of biological evolution, given enough time.

Fraction of Intelligent Communities Developing a Technical Phase, f_c. Similar remarks can be applied to the development of technical civilization. But we are somewhat hampered by the deplorable historical tendency for that civilization which is slightly ahead technologically to engulf and destroy civilizations which are not quite so advanced; as a result we have no information from local history on the likelihood of development of technical societies. European civilization has successfully destroyed much of the record for all other societies. Yet this same historical record clearly demonstrates the selective advantage of technical societies.

The prime characteristic of intelligence is its tendency to control and acquire information about its environment and to extrapolate into the future from such information. Given a minimum level of technological proficiency there would, given this line of argument, seem to be an evolutionary pressure to engage in interstellar communication. However, recognizing that many accidents have occurred to produce our technical civilization, I choose a value for f_c of 10^{-2} (or 1 percent). The reader is invited to select any other value he finds reasonable for any of these factors and to follow through to evaluate its implications.

If we momentarily withhold evaluation of L and insert the values arrived at in the preceding discussion, we have the interesting result that:

$$N = (10 \times 1 \times 1 \times 1 \times 1 \times 10^{-2})L$$
$$= (10^{-1})L$$

The number of technical civilizations is roughly equal to 10 percent of the mean lifetime in years of such civilizations. This an extremely interesting conclusion, since it means that the final value of N depends almost entirely on what value one chooses for the least well-known parameter. Let us pick two examples, labeled "pessimistic" and "optimistic."

First, the pessimistic (although perhaps also realistic) case: Reading the daily newspapers one can easily come to the conclusion that our civilization will end within a few decades. In this case, L would be on the order of 10 yr since we have acquired the capability for interstellar communication only very recently. Inserting an average value of $L = 10$ yr results in the prediction of just one technical civilization currently existing in the galaxy: *ourselves*. For those who have felt that the choices of other quantities in the equation were a bit optimistic, we now offer a sobering conclusion: Of all the conceivable planets orbiting all the 10^{11} stars in the galaxy, only one supports a technical civilization. This is a serious possibility and, if correct, a massive search for extraterrestrial intelligence will be a monumental waste of time and funds. But we have no way of *knowing* that this pessimistic prediction is correct.

The second possibility, a more optimistic one, has some merit to it, perhaps more than the first. Perhaps some small fraction, say 1 percent, of technical civilizations achieve the communicative phase only after they have successfully learned how to handle their social and political problems. We might further imagine that they decline to develop weapons of mass destruction and instruments of ecological and population catastrophes; or that they develop such devices in a way that poses no threat to their continued existence. This is a very different situation from our own, but if it is a real possibility, we can readily imagine very long lifetimes for such societies. Indeed, their lifetimes might be measured on geological or stellar evolutionary timescales. If something approaching 1 percent of all technical civilizations is able to follow this route, then the average lifetime of all technical civilizations, ranging from those that make it to those that do not, will be about 1 percent of 10^9 yr or 10^7 yr. The corresponding number of civilizations in the galaxy today then becomes roughly 1 million.

All this is, of course, purely speculative. It is especially so since we have not even one example of the lifetime of a technical society. Given the enormous significance of success in a search for extraterrestrial intelligence, it would seem well worth the effort to make such a search, particularly if we bear in mind the possibility of a million civilizations in the galaxy. A value for N of 10^6 implies a mean distance between civilizations in our region of the galaxy of a few hundred light-years. This is much less than the distance that our fledgling technology is already capable of communicating across. An examination of likely stars within a few hundred light-years would not seem out of order, nor would it be a proposition involving the expenditure of our maximum efforts.

There is a critical point contained in the equation for N. The ratio of N to L is the annual number of civilizations that are crossing the threshold of technical capability for interstellar communication. That is, N/L is the number of technical societies arising each year; it is not affected by our uncertainty in the choice of L. Based on the assumptions adopted previously, the birthrate is one civilization per decade. We arose between 10 and 20 yr ago. Thus there is not likely to be a single other civilization in the galaxy with which we could communicate that would be younger than ourselves. In the whole of the galaxy we are probably the most immature technical society. Certainly there are other societies less mature than we, but they cannot be communicated with and so are not under consideration. Thus all technical civilizations in the galaxy are in advance of *homo sapiens*, and most are considerably more advanced. Indeed, we might expect civilizations millions of years or more in advance of ourselves.

A fashionable current pastime is predicting what conditions here will be like in the year 2000, just a few decades hence. The difficulties in making these predictions are considerable. Thus we cannot even hope to speculate reasonably on the state of a civilization a million or more years ahead of us. This is the single most critical factor in the search for extraterrestrial intelligence: they are smart and we are dumb.

There are numerous questions relating to the method of searching for extraterrestrial civilizations at roughly our present technological level. If the essential technique involves a refined application of contemporary radio astronomy, the fundamental question concerns what wavelength should be used, what stars are to be examined, what bandpass and integration time constant to choose, and numerous other technical details.

However, no search can even be initiated without approaching such questions at the outset. Edward Purcell noted that the situation is rather like arranging a meeting with someone in New York City but not specifying the precise location. Let any two individuals agree to meet in New York City without specifying the location. What will be their strategy for finding one another? The obvious strategy is for one to travel to the location to which the other is most likely to travel: Grand Central Station, the top of the Empire State Building, the Statue of Liberty, Radio City Music Hall, and so on. The possibilities are numerous and must all be examined because they have essentially equal probability. But the list is not extremely long.

This example is rather similar to the difficulty of choosing which station to listen to in the radio region of the electromagnetic spectrum. There are an enormous number of distinct channels in the radio window of the earth's atmosphere. However, some channels are more likely than others to be used by extraterrestrial civilizations. Philip Morrison was the first to suggest that a natural frequency to be adopted by communicating civilizations would be the "broadcast" frequency of the neutral hydrogen gas spread between the stars of the galactic disk. This is located at a frequency of 1,420 megacycles or 21-cm wavelength. At the time of this suggestion, 1959, the 21-cm line was the only microwave spectral line that had been detected by radio astronomers. Hydrogen is by far the most abundant element in the universe.

The two nearest stars of approximately solar spectral type, Epsilon Eridani and Tau Ceti, were observed by Frank Drake in his Project Ozma during the early 1960s. Drake chose the 21-cm line and monitored these stars for several weeks. Though the attempt was unsuccessful, it is important as the first conscious effort to search for evidence of activity by extraterrestrial intelligence. (Numerous unconscious attempts are constantly being made as we will see below.)

It is very unlikely that anyone would be beaming information directly at us unless they suspected our existence and had some interest in contacting us. However, an expanding technology would seem to be characterized by increasing energy consumption. This is indeed a major aspect of the current environmental crisis. The expanding technology of a civilization much in advance of ourselves would be able to channel enormous amounts of energy into interstellar communication. N. S. Kardashev, a Soviet astrophysicist, has proposed that there are three general classes of civilization based on the amount of energy available for communication. The three types are as follows.

Type I Energy resources available for communication would be similar to those of the entire contemporary terrestrial technology (about 4×10^{19} ergs/sec or 4×10^{12} watts)

Type II A civilization capable of utilizing and channeling the total energy output of its parent star (about 4 x 10^{33} ergs/sec or 4 x 10^{26} watts)

Type III A civilization with access to the entire power output of its galaxy (about 4 x 10^{44} ergs/sec or 4 x 10^{37} watts).

The second and third types are considerably in advance of us, but if we extrapolate the energy increases we have been experiencing on the earth, the Type II phase will be reached in about 2,000 yr. The precise source of such a large quantity of energy need not concern us here; that is a matter for future technology. It is possible that such a level of expertise would not be achieved in a few thousand years or even a few hundreds of thousands of years. The central point is that we can conceive of civilizations that can control such quantities of energy. If the method for deriving and utilizing such quantities of energy exists, then it may not be unreasonable to be on the lookout for civilizations which use them.

If we admit that Type II and Type III civilizations can exist, a corresponding change in search strategy is in order. To detect a civilization controlling a sizable fraction of the energy output of a galaxy would require us to examine other galaxies rather than nearby stars. The distance problem for interstellar communication becomes much less severe a restriction in this case, and there is the distinct advantage that an entire galaxy can easily be examined with a single radio telescope. (When we attempt to examine millions of individual nearby stars to search for a Type I civilization, the search becomes very time-consuming.)

A question frequently raised is whether to send or receive. If everyone is busy searching and no one is sending, then no signals can be detected even though civilizations may be separated by a few tens of light-years. Indeed, this point is often raised as an objection to the entire concept of searching for extraterrestrial technology. We noted that our own technological activities create a signal that propagates isotropically and at the speed of light. With a radio telescope sufficiently large, eavesdropping would be easy. In our own case there is a spherical wavefront expanding at the speed of light from the solar system. This wavefront has a present-day diameter of about 40 light-years. The diameter increases, of course, by 1 light-year annually. The subject matter of this broadcast includes among other things the radio arias of Enrico Caruso, the 1930 American elections, German politics, and the Scopes evolution trial. If there is a technical civilization with even roughly our own modest level of technological capability within 40 light-years of the sun, then they may be aware of our existence. Perhaps a return broadcast is on its way here now, and we can expect to be receiving it sometime around 2010. We would not necessarily expect any eavesdroppers to be able to decipher the content of the earth's radio emission, but there are autocorrelation techniques available that would at least indicate that the signal is not of astronomical origin. For some, the question of whether to broadcast or to receive reduces to fears that allowing other civilizations to know about our existence will result in someone flying out of the sky to do us harm. Those who have succumbed to this style of thinking should be very concerned indeed, since we *have* announced ourselves, and we are continuing to announce ourselves with increasing clarity as the volume of broadcasting increases. But I do not think that their concerns are valid.

A major question is posed by the supposition that Type II or Type III civilizations exist: How is it that astronomers, in the course of pursuing their research, have not stumbled across evidence for extraterrestrial technology? Some workers in the field have been greatly concerned about this seeming paradox, and are persuaded that the number of extraterrestrial civilizations in the galaxy may be very close to the pessimistic prediction (i.e., unity). Certainly if a million civilizations are extant in the galaxy, then we would expect at least a handful of these to be of Type II, if not one of them to be of Type III. There is, in fact, a range of astronomical phenomena currently not understood that should be more closely examined. (This is not to suggest that any of those observations described below are evidence for extraterrestrial intelligence; rather, the list is an attempt to demonstrate the type of phenomenon for which we might reasonably be on the lookout.)

One example of astronomical phenomena which are not currently understood is the decameter radiation from the planet Jupiter. This radiation appears to originate in short bursts from very small areas on the planet. The locations are fixed. There is, at present, no very good theory for the origin of this radio emission, though there exist some reasonable suggestions for the modulation mechanism. Certainly this is not evidence for a Type I civilization on Jupiter, but it is equally correct to say that the evidence does not rule out such an interpretation.

A second example is the hydroxyl (OH) emission regions in the galaxy. This molecule radiates in the radio window at a wavelength of about 18 cm. Its emission seems to be highly localized, primarily surrounding the hot O and B type stars. The spectrum of this radiation is difficult to understand in terms of normal emission mechanisms. The currently accepted suggestion is that the molecule radiates much like a natural maser in interstellar space. The discoverer of the OH emission, Alan Barrett, proposed that the emission might be due to extraterrestrial intelligence. However, precisely because the O and B stars are so luminous, they are unlikely to have planets with conditions suitable for the origin and evolution of intelligence. Even more important, such stars have stable main-sequence lifetimes of only a few million to tens of millions of years.

A third example might be galactic infrared sources. These appear to be large spherical shells of matter that are absorbing starlight from inside and reradiating it at longer wavelengths in the infrared region. Perhaps these sources are stars or even planetary systems in the initial stages of formation. Our present theory of stellar and planetary evolution predicts that the earliest phases of the process should appear this way. On the other hand, Freeman Dyson has proposed that an advanced civilization might reorganize the nonstellar material of its planetary system into a large number of small bodies orbiting the parent star at roughly the same distance and in all possible orbit planes so as to make the most efficient use of the total power output from the star. (The vast majority of solar radiation is now completely wasted, with the earth intercepting only a small proportion of the total.) If the number of bodies in such orbits were sufficiently large, they might conceivably absorb virtually all the radiation emitted by the central star. The shell we would see would have a typical planetary temperature—a few hundred degrees Kelvin. However, the temperatures of the objects observed are mostly 400 to 1000°K, which may be biologically uninteresting. Yet this may be a parochial objection, if we are willing to stretch the definition of conditions suitable for life. If some of these infrared sources are extraterrestrial civilizations (and there may be ones with lower temperatures that have not yet been discovered), then they would be classified as Type II.

The classic example of an astronomical discovery seeming to imply extraterrestrial intelligence is the pulsars. One of the original speculations regarding these sources of precisely timed radio pulses was indeed "LGM"—Little Green Men (perhaps a galactic Loran system or something of that nature). Subsequent investigation has revealed the more likely interpretation that the pulsars are in fact neutron stars—a missing link in the theory of stellar evolution. Other classic examples come to mind, including Kardashev's suggestion that the quasistellar source CTA 102 was a Type III civilization.

Recently, strong infrared emission has been discovered emanating from galactic centers, including the nucleus of the Milky Way. Centered at 70 μ, this radiation has no immediately obvious interpretation. This is another observation well worth pursuing.

The point is that the search for extraterrestrial intelligence involves normal astronomical research procedures. We will be forced to wrap up areas of uncertainty in our knowledge before stating that we have not observed such civilizations and cannot hope to do so. This is a very good reason to engage in a search for extraterrestrial technology: we will need to resolve a number of unsettled problems while discovering and resolving new ones as a by-product of the search. Even if a null result is achieved, we will have vastly extended our knowledge and gained a new perspective on our significance in the universe.

Our techniques for investigating new astronomical problems are still quite primitive. For example, to investigate time variation in the recently discovered infrared source at the galactic center, it would be necessary to observe the region with a Josephson-effect detector mounted in a satellite. However, the Josephson effect was discovered just a few years ago. Our ability to detect extraterrestrial intelligence very much depends on our own technological frontiers. Also we may already have discovered extraterrestrial technology and not have correctly interpreted the evidence.

No means exist to predict science and technology ten thousand, a hundred thousand, or a million years into the future. We cannot correctly forecast what methods will be invented to investigate those questions that will arise or even have already arisen. As an example, we might reasonably question the assumption that communication must be restricted to the channels and means we have discussed. The most likely reason for us to use radio if alternative, more efficient means exist is this: the more advanced civilizations are looking for the dumb, emerging technical societies and seeking to give them information that may help them make their L a large number. This is the best rationale for a search for extraterrestrial intelligence using existing technology. And it is quite independent of the "eavesdropping" possibility we discussed above.

Even if unsuccessful, the search for extraterrestrial intelligence has a further great virtue—its almost uniquely interdisciplinary nature. There is no other subject that so completely relates the many sciences known to men. In the single equation for estimating the number of contemporary civilizations, we begin with astrophysics, progress through questions of the evolution of life, higher organisms, intelligence, and society and into subtler questions in the behavioral and political sciences. The accuracy of our knowledge of the factors of this equation begins at better than a factor of ten and melts away to an uncertainty over a range of millions or even billions. Perhaps we have not even identified all the correct variables and may even have missed some of the most important ones.

In conclusion, let us reflect on man's uniqueness in the universe. Tortuous paths have led, over some five billion years of evolution, to a technical society on this planet—an evolution marked by so many branch points that we can have no hope that a similar path has been followed elsewhere. There can be no humans beyond the earth. These thoughts are well expressed by Loren Eiseley in *The Immense Journey:*

> Lights come and go in the night sky. Men, troubled at last by the things they build, may toss in their sleep and dream bad dreams, or lie awake while the meteors whisper greenly overhead. But nowhere in all space or on a thousand worlds will there be men to share our loneliness. There may be wisdom; there may be power; somewhere across space great instruments, handled by strange, manipulative organs, may stare vainly at our floating cloud wrack, their owners yearning as we yearn. Nevertheless, in

the nature of life and in the principles of evolution we have had our answers. Of men elsewhere, and beyond, there will be none forever.[2]

Further discussion of the subject matter of this chapter can be found in "Communication with Extraterrestrial Intelligence" edited by Carl Sagan; "Intelligent Life in the Universe" by I. S. Shklovskii and Carl Sagan; "The Cosmic Connection: An Extraterrestrial Perspective" by Carl Sagan; and "A Search for Life on Earth at 100-Meter Resolution" by Carl Sagan and D. Wallace. Full information on these references can be found in the Bibliography.

[2] Loren Eiseley, "The Immense Journey" Random House, New York, 1957.

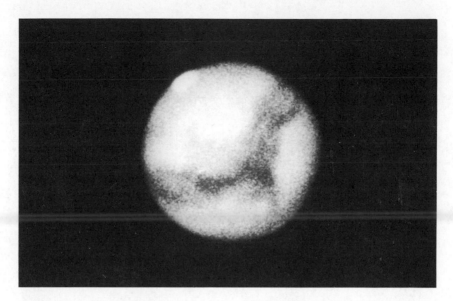

Figure A Mars showing dark areas and pole caps.

The Astronomical Background

In the preceding chapter Sagan has given us an interdisciplinary formula for determining the number of civilizations that could be expected in the galaxy. The first three factors, R_*, f_p, and n_e, are basically astronomical in character. In the next chapter, A. G. W. Cameron examines the factors f_p and n_e, the fraction of stars having planetary systems and the number of planets per planetary system ecologically suitable for the origin and evolution of life.

Present ideas concerning the origin of the solar system are extremely different from those held forty years ago. The latter ideas are described by Cameron as "dualistic," whereas the modern ideas lie in the class that he calls "monistic." Until the views concerning the origin of the solar system had changed from dualistic to monistic, so that the solar system could be thought of as a natural product of the processes of star formation, the subject of interstellar communication could not exist. Forty years ago the factor f_p was thought to be practically indistinguishable from zero. Until that opinion changed, no other speculations on the subject were worthwhile.

The quantity R_* used in Sagan's equation oversimplifies the problem of interstellar communication somewhat, because in a certain sense it views the emergence of new intelligent civilizations in the galaxy as a process that should now be going on at approximately the mean rate of the past. However, in the next paper Cameron indicates that most of the sites of probable extraterrestrial intelligent civilizations were established many billions of years before the formation of the solar system, and, therefore, most galactic communities, if they still exist, will be ancient, and hopefully, much wiser than ourselves.

2

Planetary Systems in the Galaxy

A.G.W. Cameron   *Harvard University*

Requirements for Planetary Systems

A principal requirement for the existence of an extraterrestrial civilization is a suitable environment. This means both a suitable planet and a suitable star about which that planet revolves.

An important environmental factor to consider is the question of planetary temperature. A planet acquires that temperature at its surface which represents a balance between the arrival of heat from the central star and the reradiation of that heat into outer space. The farther away the planet from the central star, the colder it will be. In the solar system, the earth is obviously well suited with respect to temperature. Over large parts of its surface, the temperature usually exceeds the melting point of ice. Nowhere on the earth does the temperature ever reach the boiling point of water. These are the conditions best suited for life as we know it.

If the earth were moved inwards in the solar system, closer to the orbit of the planet Venus, it would become hotter. Calculations indicate that the earth could not be moved all the way into the orbit of Venus and still remain suitable for life. Under present circumstances, a good part of the incoming sunlight is absorbed by the surface of the earth and reradiated in the infrared; much of the infrared radiation has difficulty escaping through the atmosphere because of absorption by water vapor and other triatomic molecules. This method of heating the surface is known as the greenhouse effect: in a greenhouse, sunlight enters easily, but the panels in the greenhouse make it difficult for the energy brought in by the sunlight to escape again. This results in raising the temperature in the interior of the greenhouse, or on the surface of the earth, above the level which would exist if all of the received energy could easily be reradiated. As the earth moves closer to the orbit of Venus, the greenhouse effect is expected to become more important, so that the temperature in the atmosphere would rise rapidly. As the temperature rises, more water is evaporated from oceans and lakes and rivers, thus increasing the number of infrared absorbers in the atmosphere, and hence augmenting the greenhouse effect. Very soon the balance would be tipped, and the temperature in the atmosphere would become so high that most of the water in the oceans would evaporate to form a huge, dense steam atmosphere. It is difficult to imagine life as we know it existing under these circumstances.

We do not yet know whether there is life on Mars. The fact that we are uncertain about this point is an indication that the earth might be moved outwards in the solar system about as far as the planet Mars, without necessarily destroying all of the life upon its surface. Very much more of the earth's surface would fall to a temperature below the freezing point of water, making life much more marginal over most of the planet. Undoubtedly some forms of life would survive under these circumstances; perhaps a more relevant question would be whether life could originate in the first place under these circumstances.

Thus there is a significant span of radial distance in the solar system in which the temperature at a planetary surface like that of the earth is suitable for the existence of life as we know it. Within the solar system, the orbits of the planets increase in radial distance in an orderly fashion, as do the orbits of some of the regular satellites about the major planets in the solar system. This may be a fairly universal characteristic of planetary systems. In that case, temperature conditions suitable for the existence of life as we know it are likely to prevail upon at least one, but not more than two, of the planetary bodies in any stellar planetary system.

Another environmental requirement for extraterrestrial life is a long-term stable set of temperature conditions. It has taken life on earth approximately 4.6 billion years to evolve to the point where man has emerged. This time period might be more or less on another planet revolving about another star to permit an intelligent species equivalent to man to emerge, but the time is nevertheless likely to be a few billion years. Even during the lifetime of the earth, the temperature may have risen significantly; studies of stellar evolution tend to indicate that the luminosity of the sun has increased by a few tens of percent during the history of the solar system. Such studies also indicate that the sun may live for approximately another 5 billion years in much its present condition, but with its luminosity increasing a few more tens of percent, before it swells up to become a red-giant star, with a rapid and dramatic increase in its luminosity, sufficient to destroy all life upon the surface of the earth.

The apparent need for a long evolutionary lifetime before an intelligent species can emerge upon the surface of a planet places some important restrictions on the type of star about which one might expect to find such an intelligent species. A star which is only some 15 percent more massive than the sun, when initially formed, is expected to last only some 5 billion years before swelling up to become a red-giant star, with a corresponding large increase in luminosity. An intelligent species near such a star might well have time to evolve, but shortly thereafter it will face a severe environmental crisis. Still more massive stars have even shorter lifetimes; the most massive stars observed in space live only a few million years before they become red-giant stars. Planets may be formed in orbit about such stars, but it seems very unlikely that intelligent species, based upon carbon organic chemistry, will have time to evolve.

Thus it appears that we should not look for planetary systems containing intelligent species about stars much more luminous than the sun. In the other direction, toward lower luminosity and lower stellar mass, no lifetime restriction exists. Such stars maintain a quite constant luminosity over periods longer than the present age of the universe. We shall have more to say about the character of the planets which may contain life in such stellar systems later on.

In addition, there is another type of environmental stability which is very important to an intelligent species: stability of the planetary orbit. In general, this makes binary-star systems unattractive environments for life-bearing planets. The majority of the stars in space appear to be members of stellar binary systems. The two members of a binary pair can have a wide variety of orbits. However, there are some systematic regularities in these orbits.

Some of the binary stars orbit very close to each other, with separation distances small compared to the orbital radius of Mercury. These orbits are quite circular. Life-bearing planets well outside the mutual orbit of the two stars might bring forth intelligence if both members of the binary pair have masses comparable to that of the sun or less, so that their stellar evolutionary lifetimes are sufficiently long. Climatic conditions on such planets may well be considerably more variable than those on the earth, but they are probably endurable by intelligent species. Indeed, under some conditions, they might even stimulate the evolution of such species.

As the separation distance between the stars in a binary pair increases, the orbits cease to be circular. When the separation distance exceeds the orbital radius of Mercury, the mutual orbit of the two stars becomes appreciably elliptical, and as the separation distance increases, the eccentricity of the ellipse increases. Binary stars with very large separations, amounting to many tens of astronomical units (an astronomical unit is the distance from the earth to the sun) tend to have exceedingly elliptical orbits. Such stars, in their motion about each other, undergo large variations in their separation, from a maximum distance which may be many tens of astronomical units to a minimum distance of the order of an astronomical unit or less. This type of motion is quite inconsistent with the presence of planets in stable orbits about either of the stars. Such planets would be quickly kicked out into outer space as a result of mutual perturbations by the motions of the stars, provided that they were not swallowed up by one of the stars in the course of a close passage. In general, most binary systems, and all binary systems with large separations between the components, can be regarded as unlikely locations for life-bearing planets.

Thus we must direct our attention, in general, to single stars not much more luminous than the sun. We have a further requirement to make on such stars: they should have a decent content of heavy elements. One of the major areas of recent advance in our knowledge of the evolution of stars is the appreciation of the role of stars in the manufacture of heavy elements. We shall discuss this in more detail later but, in essence, the discovery of stellar nucleosynthesis has shown that all of the common elements which exist upon the earth, with the exception of hydrogen, are formed as a result of nuclear reactions that take place during the evolution of stars, mostly in the late stages of evolution of fairly massive stars. Thus in recent years a general explanation has been proposed. In this view the galaxy may have consisted of gases containing only hydrogen and helium when it was first formed; stars formed in the galaxy and started to manufacture heavy elements, and, gradually, the level of abundances of the heavy elements in the interstellar gases increased. As stars formed during the history of the galaxy, the gases involved in the formation of the oldest stars would be greatly depleted in heavy elements, and hence the formation out of heavy elements of planets about such stars would be greatly impeded. However, today it appears that this is much less of a restriction than it formerly seemed to be, since even the oldest stars seem to have acquired a significant content of heavy elements very early in the history of the galaxy, and perhaps even before the galaxy was formed. These considerations are important for making estimates of the oldest stars which may contain intelligent species in our galaxy. The age of the sun appears to be between one-half and one-third of the age of the galaxy, and hence many of the stars in the galaxy may have had conditions suitable for the emergence of intelligent species for a period of time longer than the age of the solar system itself. Such considerations must obviously play an important role in our considerations when we try to decide the optimum approach which we should make toward searching for interstellar communications.

In the remainder of this chapter we shall outline some ideas of the expected history of the universe as it is viewed in current astrophysical thinking, and we shall consider when elements were formed, how galaxies were formed, and how stars in planetary systems are formed within our own galaxy. We shall then try to discuss briefly the prevalence of planetary systems suitable for intelligent species within our galaxy.

Cosmology

The basic structure and evolution of the universe, generally called cosmology, has been a subject extensively debated in recent years. Many different hypotheses concerning conditions in the early history of the universe have been put forward. However, a few years ago it was discovered that there is a background radiation in the microwave region of the spectrum in which the earth is continually bathed; this radiation is incident on the earth uniformly from all directions. Most astronomers believe that this background radiation is a relic of a former hot state in the universe; as the universe has expanded, the radiation has expanded with it, and the photons in the radiation have shifted toward lower and lower energies, with their wavelengths shifting far toward the "red" end of the spectrum. In fact, the expansion has been so great that the photons have shifted well beyond the ordinary red end of the visible spectrum, and are now far out into the radio region. This is simply a manifestation of the doppler effect, according to which objects receding from us at a rapid rate have their characteristic spectral lines redshifted. In general, the farther an object is from us in the universe, the greater its velocity of recession from us, and the greater the redshift of its photons will be. The sources of the microwave background radiation are very far from us indeed in space and time, and hence their redshift is especially great.

If the universe is expanding from an initially hot and dense state, then this supports some form of "big-bang" cosmology. Hence cosmologies of this type are the only ones that we will consider here.

At present, one of the most intriguing problems in cosmology is the question of whether the universe is symmetric, that is, whether it contains equal amounts of matter and antimatter. Matter and antimatter annihilate when they come into contact with each other, forming various types of radiation as annihilation by-products. Among the most important of these forms of radiation are hard gamma rays. However, we see so little gamma-ray background impinging on the earth that we can be quite certain that there is no significant amount of antimatter in our own galaxy. It is generally believed that this statement can logically be extended to all of the galaxies in our local cluster of galaxies. However, the question of whether there may be matter in some clusters of galaxies and antimatter in other clusters of galaxies is entirely open. This would constitute a symmetric universe, a concept which is very appealing to many physicists.

If the universe is not symmetric, but contains only matter, then early in its history some events occurred which have some significance to our story. When such a universe is very young and has a very high temperature, then ordinary atoms and their nuclei cannot exist, because the intense radiation fields at those high temperatures would tear them apart. Only the basic neutrons and protons can exist. However, as the universe expands, it cools, and a point is reached at which the neutrons and protons can combine with one another to form helium nuclei. Spectroscopic studies have shown that the stars in our galaxy and the matter in many other galaxies have a remarkably uniform content of helium, about 1/4 of the mass. It is a striking feature of nonsymmetric cosmologies that they predict that just about this amount of helium will be produced by this combination of neutrons and protons in the early history of the expansion of the universe.

On the other hand, in a symmetric universe, as has recently been postulated by Omnes, patches of matter and antimatter are formed at a very early stage, which then diffuse together, so that most of the material is mutually annihilated. As time goes on, the remaining matter and antimatter patches tend to collide with one another, so that there is annihilation and a bounce when a matter patch meets an antimatter patch, but merging together when a matter patch meets another matter patch or when two antimatter patches meet. In this way, the surviving matter grows in mass by merger until masses of the order of clusters of galaxies have gathered together. A similar process takes place with patches of antimatter. One of the features of these processes is that the neutrons, which might be combined with protons to form helium, are selectively lost through faster diffusion and subsequent annihilation with antineutrons, and hence practically no helium can be formed in such a symmetric cosmology. Nevertheless, the fact that even very old stars seem to have about as much helium as the sun, which was formed much later in the history of our galaxy, argues that the transformation of 1/4 of the mass of the universe into helium (and presumably antihelium) must have taken place at some pregalactic stage. This is one possible indication that a stage of pregalactic nucleosynthesis may be required.

As the universe expands, initially the radiation and the matter (and antimatter) expand rather uniformly within it. However, a stage is eventually reached at which the matter can start clumping together and fragmenting into isolated clouds of gas. These clouds of gas will continue expanding for awhile, but will reach a maximum stage of expansion, and will then fall back together toward common gravitating centers.

A few years ago it was generally assumed that the gas clouds would keep on collapsing until they formed galaxies. However, there is now reason to doubt that galaxies can be the direct products of this collapse.

Consider the structure of our own galaxy. It has a nucleus containing a dense concentration, spherically distributed, of very old stars. The system of very old stars extends well beyond this nucleus, remaining roughly spherically symmetric and extending to very large distances, many tens of thousands of light-years, away from the galactic center. This outer structure is called the galactic halo. The gas in the galaxy and all of the newer stars are concentrated in a thin disk, where the gas and stellar motions have essentially a circular direction around the center of the galaxy. The most prominent features in this flattened disk are the spiral arms, which extend outwards from near the galactic center and are loosely wrapped around it, and in which are concentrated much of the gas and most of the young hot stars. The sun is about 30,000 light-years away from the center of the galaxy and lies in the galactic disk.

About 3/4 of the mass of the sun is composed of hydrogen, 1/4 of helium, and all of the rest of the elements constitute only between 1 and 2 percent of the mass of the sun. The elements that make up the interior of the earth constitute only about 0.3 percent of the mass of the sun. It is these heavier elements with which we are thus particularly concerned in the stellar compositions throughout the galaxy.

Most of the stars in the galactic disk appear to have about the same fraction of heavy elements as does the sun. Even very old stars, formed near the beginning of the life of the galaxy and still gathered together in galactic clusters, often seem to have about the same heavy element content as the sun. However, typical stars in the galactic halo are depleted in heavy elements relative to the sun, but even these stars have approximately 1/3 as many of them as does the sun. In principle this is still an ample amount of heavy elements to form planets suitable for life-bearing species. A rare few of the halo stars are, however, extremely highly depleted in heavy elements relative to the sun, with depletion factors ranging from 10^{-2} to 10^{-3}. These are certainly not likely to have habitable planetary systems, but the number of these stars appears to be very small.

It is generally thought that the galactic halo stars were formed during the general collapse of gas which formed the galaxy. This seems very plausible, but it immediately raises an important issue: how did these galactic halo stars acquire their abundances of the heavy elements? Once again, a stage of pregalactic nucleosynthesis seems to be required, but in this case to supply the heavy elements in the galactic halo stars. This is an argument which seems to require a pregalactic stage of nucleosynthesis independent of whether the universe is symmetric or not.

Because this idea is relatively new, it has not yet been fully developed. However, we can imagine a scenario something like the following. After the clouds of gas in the expanding universe reach their maximum degree of expansion and start to recollapse, they are likely to fragment into very massive stars, of the order of a thousand times the mass of the sun or greater. In the evolution of these massive stars, nuclear reactions create the heavy elements and possibly the large amounts of helium that are required in a symmetric universe. The gas that is ejected from these stars, enriched in heavier elements, together with the gas which did not get into the massive stars in the first place, can then gather together and collapse to form galaxies. The galactic halo stars formed during the collapse of the gas will thus contain a significant content of heavy elements. However, because the heavy elements will be somewhat inhomogeneously mixed into the gas, a few of the stars which are formed will be highly depleted in heavy elements. The old stars formed in the galactic core may have obtained most of their content of heavy elements in this same way.

We thus reach the following very important tentative conclusion. Nearly all of the single stars that have ever been formed in the galaxy could have been suitable abodes for planets containing intelligent species.

Formation and Death of Stars

We are now in a position to outline in a qualitative way the evolution of our galaxy. We start with a huge cloud of gas which is collapsing toward a common gravitating center. This cloud of gas consists of about 3/4 by mass of hydrogen, about 1/4 by mass of helium, and perhaps 0.5 percent by mass of heavier elements. In a few places, the heavy element admixture will be miniscule compared to this quantity, because of inhomogeneities in mixing into the collapsing gas. There will be density fluctuations in this infalling gas, and in some of the regions of greater density stars will form before the collapse is complete. These stars are formed at large distances from the center of the galaxy, and with components of motion directed generally toward the center of the galaxy; subsequently they will remain in orbits within the galaxy which take them from close to the center of the galaxy to large distances away from it, and their distribution about the center of the galaxy will be roughly spherically symmetrical. These are the galactic halo stars. There are more stars formed per unit volume near the center of the galaxy than far away from it, so that the halo stars exhibit a strong concentration toward the center of the galaxy.

Not all the gas which is collapsing toward the galactic center goes promptly into stars. Furthermore, this gas will have a considerable amount of angular momentum, or tendency to spin, about the galactic center. Hence the gas cannot approach too closely to the galactic center, but must settle into a thin gaseous disk which rotates around it. This gas will then give the galaxy the characteristics which allow it to be described as a spiral galaxy, as we shall see shortly. In some other galaxies, the amount of gas going into star formation is very much greater, so practically no disk is formed; we call these elliptical galaxies. In still other galaxies, the motion of the gas is very much more random, and the amount of prompt-star formation is relatively small; we call these irregular galaxies.

Once a large amount of gas has settled into the thin galactic disk, star formation should occur quite promptly within it. These stars will lie in the plane of the disk and will have orbits which are approximately circular around the center of the galaxy. There is a tendency for a temporary clustering of these stars in spiral arms; this clustering represents a departure from the purely circular motion about the galactic center.

Spiral arms are presently thought to represent density waves which travel around the center of the galaxy, propagating through the stars in the galactic disk. Suppose that such a density wave exists and that it represents a temporary clustering of stars in the disk into a spiral arm. Now let us focus our attention on one of the stars which is rotating around the center of the galaxy. The motion of such a star in a region between two spiral arms is approximately circular. When such a star enters into a spiral arm, it feels the attraction of the excess concentration of other stars in its vicinity, and its motion is modified in such a way that it spends an extra amount of time within the region of the spiral arm. Eventually it emerges from one spiral arm and proceeds toward the next one. While it is in the spiral arm for its unusually long residence time, it has contributed to the gravitational field which tends to keep other stars in the spiral arm for an unusually long residence time. It is probable that stars spend about half of their orbital lifetimes within spiral arms and the other half of their orbital life traversing the space between the spiral arms.

The presence of the spiral arms also has important effects upon the remaining interstellar gas which has not formed into stars. When the gas leaves the interarm region and enters a spiral arm, the extra gravitational attraction of the stars in the spiral arm slows down the gas and compresses it. Ordinarily, such gas has internal-pressure forces, tending to make it expand, which are large compared to self-gravitational forces which might draw parts of the gas together into a collapse toward much higher density. However, when compression takes place in the spiral arm, there is a much greater chance that density fluctuations can arise in at least a few places within the gas which are sufficiently great so that the self-gravitational forces of the gas can induce gravitational collapse within it. This surmise receives observational support, since all of the new stars which are formed in the galaxy appear to be formed within spiral arms. If one looks at another spiral galaxy, the hot, bright, highly-luminous stars are all observed to lie in the spiral arms; they burn up their nuclear-fuel resources at such a prodigious rate that their lifetime is shorter than the time required for them to escape out of the spiral arms into the interarm region.

It appears that in order for a dense cloud of the interstellar gas to start collapsing in a spiral arm, a mass several thousand times that of the sun is required in the gas. As the gas collapses, irregularities in its density and structure tend to become magnified so that the collapsing gas tends to subdivide into smaller pieces which independently fall toward the common gravitating center. This is a progressive process; each of the pieces in turn will have irregularities in structure and density, and these in their turn will grow, thus further subdividing the fragments of the gas cloud into smaller and smaller pieces. In this way the collapsing cloud becomes broken up into a large number of units from which stars will form.

Let us skip over the final details of the star-formation process, which are of interest for the formation of planetary systems, and discuss the evolutionary lifetime of a star once it has been formed. Stars derive most of the energy which they emit into space from the conversion of their internal hydrogen fuel into helium by a variety of thermonuclear reactions which take place in the center of the star. During this central hydrogen conversion process, the stars are said to be "main-sequence" stars. The hydrogen fuel keeps a star like the sun shining in much its present state for about 10 billion years. Stars much more massive than the sun, containing several tens of solar masses, may require only about 3 million years to burn out their central hydrogen into helium. Stars much less massive than the sun have very low luminosity, and hence they will remain as main-sequence stars for many multiples of the lifetime of the universe.

After a star has burned out its central hydrogen, its behavior becomes quite complex, and its evolution is very much speeded up. The helium, in its turn, can be burned to form carbon and oxygen, and in massive stars these two fuels then lead to the formation of still heavier elements when they, in their turn, burn at higher stellar internal temperatures. Stars of about four times the mass of the sun and less tend to shed much of their outer layers of gas and to settle down as compact, low-luminosity objects called white dwarfs. More massive stars tend to undergo very spectacular nuclear explosions, called supernovae, in the course of which their outer layers are violently flung into space and a small remnant, called a neutron star, remains at the center. The neutron stars emit narrow beams of radio waves as they rotate, which appear as pulses of radiation when they sweep across the earth, from whence such neutron stars are called pulsars.

Much of the gas which is thrown off in supernova explosions is highly enriched in heavier elements which have been manufactured by nuclear reactions taking place during the explosion. These heavy elements enrich the interstellar gas which remains in the galaxy so that subsequent generations of stars, when they are formed, will have a higher heavy element content than those which preceded them. The heavy element content of the sun is about 1.5 percent by mass, considerably greater than the mass fraction of heavy elements that we estimated for the initial stars condensing in the galaxy. However, it appears that most stars in the galaxy were formed very early in the galactic lifetime so that most of the enrichment of the interstellar medium in the heavy elements occurred very early. It is even possible that there has been a slight decrease in the heavy element content of the interstellar gas in recent billions of years, if there is still a significant rate of infall of gas into the galaxy from intergalactic space, and if this gas has the lower heavy element content which is characteristic of the initial galaxy. The dilution of heavy elements that occurs in this way may not quite be offset by the formation of new heavy elements in recent supernova explosions.

There are two general ways in which the age of the galaxy has been estimated. One of these relies on comparison of theory and observation concerning the evolution of the stars; the distribution of the brightness and colors of the stars in old galactic and globular clusters gradually changes as a function of time, and these changes can be computed in the theory of stellar evolution. In this way the galaxy has been estimated to be 10 to 15 billion years old.

The other method of estimating the age of the galaxy depends on radioactivities which are formed as a result of stellar evolution, and which have lifetimes sufficiently long so that the parent radioactive nuclei are still present upon the earth. Of special importance in this respect are the heavy elements thorium and uranium. The relative abundances of the thorium and uranium nuclei change as a function of galactic age, and by noting the abundances which exist in the solar system, we conclude that the galaxy is about 12 or 13 billion years old. Thus a majority of the stars and planetary systems which should exist in the galaxy can be expected to be more than 10 billion years old. For comparison, the solar system was formed only 4.6 billion years ago.

Formation of the Solar System

Let us now return to the details of the final stages of star formation which we skipped over in our preceding discussion. We only have evidence concerning those details for one system in our galaxy: the solar system. Therefore we must consider what has been concluded about the origin of the solar system.

This problem has been discussed in a scientific manner for more than three centuries, since the time of Descartes. During most of this long period, the scientific discussion has been greatly hindered by lack of observations pertaining to the very early history of the solar system. The oldest rocks found upon the face of the earth are less than 4 billion years old, so that the first several hundred million years of terrestrial history have been lost from the geological record. This is an indication of the great violence which the earth presumably underwent in its earliest times. Thus the earlier scientists had only regularities in the orbits and motions of the bodies of the solar system available for their speculations about the origin of the solar system. During the last two decades the amount of material available to guide these scientific speculations has immensely increased. This increase has occurred first, as a result of detailed scientific analysis of the properties of meteorites, which are small fragments of solid bodies believed to be part of the asteroid belt, and more recently as a result of space probes which have investigated the properties of the nearer planets and the moon. This new information has required a great deal of additional scientific sophistication in speculations about the origin of the solar system.

At the present time most solar-system scientists are in agreement that the solar system originated from a rotating flat disk of gas which they call the primitive solar nebula. This rotating disk of gas should be formed when gas collapses toward a common gravitating center. However, since it contains angular momentum, it must settle into rotation about the common gravitating center, much as did the gas which forms the galactic disk in a spiral galaxy. However, there is considerable lack of agreement on the distribution of mass in the primitive solar nebula.

One approach leads to what we may call the minimum solar nebula. This approach is based on the assumption that the collection of condensed solid material to form the planets has been very efficient so that the mass of the solar nebula is obtained from the masses of the planets by adding the missing volatile material that exists in the solar composition. For example, the rocky material which constitutes the inner terrestrial planets represents only about 0.3 percent of the mass of solar material. The two largest planets, Jupiter and Saturn, can be considered perhaps to be very like the sun in composition so that the proponents of minimum solar-nebula theories have not suggested that any correction be made for these planets. A correction is, however, needed for Uranus and Neptune, 80 percent of whose mass may consist of the heavier elements other than hydrogen and helium, which constitute about 1.5 percent of the mass of the sun. When all of this addition has taken place, the mass of the resulting solar nebula is only a few percent of that of the sun. The proponents of minimum solar-nebula theories are therefore required to assume that the sun formed directly from the collapse of interstellar gas, and that only a tiny fraction of this gas had enough angular momentum to go into orbit about the sun. The total angular momentum implied by these assumptions is very much less than that which would be expected in a fragment of a collapsing interstellar cloud. No detailed mechanisms have been suggested by which this large loss of angular momentum might occur.

At the other end of the scale we have what might be called the massive solar-nebula theories. In this case it is not assumed that the characteristic angular momentum of the fragment of the collapsing interstellar cloud is lost so that the entire gaseous mass settles down into the flat rotating disk, and there is no central star formed on the spin axis of the nebula. In a theory which involves a massive solar nebula, the sun must form from the nebula by an inward flow of mass resulting from gaseous-dissipation processes, and most of the condensed solid material in the nebula must be transported into the sun with the gas, which will occur if the solids are in relatively finely-divided form. It is necessary to assume that the massive solar nebula contained considerably more mass than is presently in the sun, probably about twice as much mass. This is explained by the fact that (1) not all of the initial solar nebula can become part of the sun because a part of the mass must be left behind in order to take up the angular momentum lost by the mass which flows into the sun, and (2) the loss of mass by the early sun in its T Tauri phase must be allowed for. T Tauri stars are young stars which have not yet started to burn hydrogen into helium in their central regions. Such stars are observed to be losing mass at a prodigious rate, something of the order of one-millionth of a solar mass per year. It must be anticipated that the sun will have lost several tenths of a solar mass by means of these primitive vigorous solar winds.

Some idea of the complexities of the massive primitive solar-nebula model may be judged from Fig. 2.1, which shows a cross section through a model which was constructed by the author in collaboration with M.R. Pine. Note that the scales in both the vertical and radial directions in this figure are logarithmic, with the distance measured in astronomical units. Counting upwards, the three lines which pass through the complete radial distance in the nebula represent the height above midplane at which (1) 50 percent of all of the mass between midplane and top of the nebula is contained, (2) 90 percent of this mass is contained, and (3) nearly all of the mass is contained and the pressure has dropped to one-millionth of the value at midplane. The other line drawn over part of the radial distance represents the "photosphere," the surface from which radiation is emitted into space, similar to the photospheric surface of the sun from which its radiation into space takes place.

The two shaded regions in the figure show parts of the primitive solar nebula which are convective, in which there are large mass motions of the gas which help to transport energy to the photosphere of the nebula from which it can be radiated into space. The convection zone in the central region of the nebula is a technical consequence of the fact that hydrogen molecules are becoming dissociated into atoms in that region. The gas in the central region is too hot for condensed solids to exist. The convection zone lying farther out is a technical consequence of the

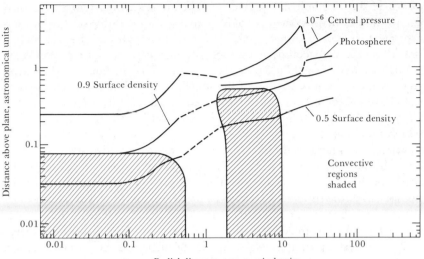

Radial distance, astronomical units

Figure 2.1 A cross-section through a model of a massive primitive solar nebula. Both axes have logarithmic scales. Convective motions take place in the shaded regions. Emission of radiation into outer space takes place from the photosphere, which is shown in the outer parts of the model; the photospheric surface coincides with the top of the innermost convective zone and is not well defined in the intermediate region where the lines are dashed.

presence of metallic iron in the condensed solids which do exist in that region, where the temperature is considerably cooler. At still larger radial distances the gas densities are too small for convection to occur.

Such a massive solar nebula should undergo a rather rapid dissipation. It has been estimated that the major part of the gas will dissipate to form the sun within only some thousands of years. Once the sun has formed, the T Tauri phase of mass loss should begin, and the remaining gas in the solar nebula should then be blown away into space. Because the gas in the nebula has a highly beneficial role in promoting the accumulation of solids into larger bodies, we must therefore ask whether it is possible that the planetary bodies can have formed in a time of only a few thousand years.

Let us discuss these accumulation processes, starting from the outer regions of the solar nebula and working inward. However, let us first consider what may happen to form the primitive solar nebula in the fragment of the collapsing interstellar cloud which is still in its collapse stage. The interstellar gas at this time is very cold, and nearly all of the nonvolatile gases are condensed in the form of small grains, having dimensions comparable to about one micron. As the density of the gas increases during the collapse, the interstellar grains start to bump into each other, and since they probably have a rather fluffy structure, they are likely to stick together upon collision. Aggregates of the interstellar grains continue to collide with each other, with a resulting growth of condensed bodies within the collapsing gas. The author has estimated that, at the time of formation of the solar nebula, these condensed bodies may have grown to about one foot in radius. This is only a typical dimension; actually there should be a large variation in the sizes of the bodies which are formed.

When the solar nebula has formed, the pressure within the gas keeps even the outer and cooler parts of the solar nebula extended to semithicknesses of the order of one astronomical unit or more. However, the solid bodies cannot be supported by the gas pressure, and they will fall through the gas and concentrate in the vicinity of the central plane of the nebula. This concentration makes it easier for them to collide with each other.

There is also a partial pressure support of the gas in the radial direction so that the gas is in balance between the force of gravity, which is directed toward the spin axis of the nebula, and the pressure in the gas and the centrifugal force due to its circular motion, both of which are directed away from the spin axis of the nebula. However, the solid bodies are not supported by the radial pressure of the gas any more than they are supported by the vertical pressure, and hence their motion must be governed by a balance between the force of gravity, directed inwards, and the centrifugal force, directed outwards. Because a greater centrifugal force is needed for the condensed solids than for the gas, it follows that the orbital velocity of the solids is greater, and hence they move relatively rapidly through the gas. The smaller bodies cannot move too rapidly through the gas because of the effects of gas drag on them, but the larger bodies will move considerably more rapidly. Thus larger bodies will be continually bumping into smaller bodies, which are moving more slowly with respect to the gas, and in this way growth toward larger sizes is very rapid. The author has estimated that planets such as Uranus and Neptune can indeed be accumulated in a time of a few thousand years.

As we move inward in the solar nebula, we reach a part of the nebula where the temperature is sufficiently high so that ice, solid ammonia, and solid methane will all have been evaporated away from the condensed bodies. This leaves only rocky material. This rocky material constitutes only about 1/15 of the volume of the colder condensed solids, and hence will initially form only a very fragile and filamentary framework. As collisions take place among the solids, presumably some combination of fragmentation and compaction and accumulation will occur. The density of solid material near the central plane of the nebula may become sufficiently high so that there will be a gravitational collection of the solids into sizable solid bodies. These will then grow by collisions with smaller bodies and by collisions among themselves. Because of the considerably greater density of material at these radial distances, it is probable that these solids can grow into bodies much more massive than the earth in a time of several thousand years. When they reach such a large size, then the surrounding gases of the solar nebula become highly concentrated toward the solid bodies by gravitational attraction, and eventually the time comes when there is a violent dynamical collapse of these gases onto the solid planetary cores. This is probably the manner in which the massive giant planets, Jupiter and Saturn, were formed.

As we move still farther inward in the solar nebula, we come to a region where the gas is still hotter, and if conditions were otherwise the same, we would expect the planets formed to become progressively more massive gas giants. However, we have now reached the region of the convection zone, in which gaseous motions occur quite vigorously within the gas and prevent a concentration of solid materials very close to the central plane of the nebula. This dispersal of the solids into the vertical direction greatly interferes with planetary-accumulation processes, so that in the time scale of a few thousand years the amount of material which can collect together to form planets is very much smaller than in the region of Jupiter and Saturn. Nevertheless, this is the region where we should expect the formation of the inner terrestrial planets. These solid bodies would have been too small to have acquired much of a primitive atmosphere of solar nebula gases, and the T Tauri phase solar wind is probably sufficiently vigorous to eliminate all of the primitive atmosphere drawn from the primitive solar nebula, thus allowing a secondary atmosphere to arise only after the vigor of the T Tauri phase solar wind has diminished.

When a planet such as the earth is formed under these conditions, the time of formation is so short that a great deal of the gravitational potential energy released by bringing the planetary mass together will be retained as internal heat. Thus much of the planet initially will not only be molten, but indeed much of the rocky material will have been vaporized when it landed upon the surface of the earth. However, this vapor should all have recondensed onto the planetary surface in a time of some few thousands of years so that the initial earth then would be molten throughout.

Within the earth the metallic iron, being of greater density, will have separated out and descended to the center to form the core of the earth. It will have taken with it certain other elements such as nickel, alloyed with the iron, and sulfur in the form of iron sulfide. The outer part of the earth, composed mostly of magnesium silicates and probably containing very little in the way of volatile elements, will form an overlying mantle. As time goes by, the radiation of the internal energy of the earth from the surface into space where the energy is transported by convective motions in the molten rocks, can gradually cool the earth and allow the mantle to solidify from the bottom towards the surface. If the rate of loss of energy from the primitive earth is retarded through the acquisition of a secondary atmosphere on the earth, which would provide a lower temperature-radiating surface, then the solidification of the surface of the earth may be postponed for several hundreds of millions of years after the formation of the earth.

When the final solar nebular gases are removed by the T Tauri solar wind, the inner solar system will contain a great deal of smaller solid debris which has never been raised to a very high temperature. This material will contain volatile elements in significant abundance, and the earth will sweep up a great deal of this debris, possibly acquiring the final 10 percent of its mass in this manner. This will be the source of the more volatile solid materials within the earth and also of the atmosphere and the oceans.

Thus we can think of the history of our planet as consisting first of the formation of a molten rocky system, swept bare by the violent gales within the T Tauri solar wind. Onto this desolate fiery lava rains a great deal of additional rocky chunks from outer space, the rain gradually petering off over a hundred million years or so of time. The solar wind diminishes in force, the earth starts generating a magnetic field, and protected by this field from the solar wind, a secondary atmosphere can form by outgassing of the volatile elements brought in by the late rain of rocky debris. The gases will include water vapor, carbon dioxide, nitrogen, and smaller amounts of ammonia and methane or other volatile organic compounds. The water condenses to form the primitive oceans, which may initially cover all of the land surfaces of the earth. However, material of lower melting point and lower density starts to emerge through cracks in the primitive crust, forming a large initial continental mass. The initial crustal surface is set into motion by moving masses of rock within the earth, and the large continental mass is broken up into smaller pieces which are transported around the earth, forming the present continents. This motion is still continuing and is called continental drift.

Meanwhile, organic evolution is taking place within the atmosphere and ocean, as more and more complex organic molecules are formed as a result of heat, radioactivity, electrical discharges, and perhaps ultraviolet radiation from the sun, all of which can manufacture amino acids and probably more complicated molecules from the mixture of primitive gases in the atmosphere. These products concentrate in the oceans, and the stage is set for the origin of life upon the earth.

Generalities of Planetary Systems

The foregoing remarks concerning the origin of the solar system can be generalized to some extent in a discussion of the probable character of other planetary systems within the galaxy. We have remarked previously that even most of the early stars formed in the galaxy probably have about ⅓ of the mass fraction of heavy elements as the solar system. While this makes the planetary-accumulation process somewhat less efficient, it probably allows large numbers of systems to be formed with planets having essentially earthlike characteristics. Essentially all modern theories of formation of the solar system indicate that the processes should take place as a natural consequence of the formation of the sun, and therefore, by extension, such theories indicate that most single stars in the galaxy should be accompanied by planetary systems.

The story which we have outlined above indicates that we should expect general architectural similarities among these planetary systems. In the inner region of the planetary system, corresponding to the presence of a convection zone, planets of the terrestrial type should be formed. Farther out there should be gas giants, and these giants are probably somewhat variable in mass, depending on the amount of gas in the primitive gaseous disk available to participate in the dynamic collapse upon large planetary cores. Several stars near the solar system have been found to be accompanied by massive bodies having masses of the order of Jupiter or greater, and these have probably been formed in this manner. Still farther out may be planets made mostly of rock and ice, much like Uranus and Neptune, although perhaps with less or more of the gases from the nebula in their atmospheres.

However, some surprising conditions can occur in planetary systems associated with low-mass stars. The theory of the massive solar nebula predicts that the composition of the planets which will have a terrestrial temperature range maintained at their surface will be significantly different from that of the earth in such systems. This is so because the composition of the planets which form in a primitive solar nebula depends on the initial temperature in the nebula, which results from the compression of the gas as it collapses from interstellar space. But the temperature maintained on the surface is a function of the mass of the star formed at the center of the system, and the luminous output of energy from such a star varies about as the fifth power of its mass. Thus, according to the generalization of the theory of the massive solar nebula, in a low-mass stellar system, the luminous output is very low so that terrestrial temperature conditions are only achieved in an orbit very close to the star. However, the temperature of the gaseous disk even at such a close distance should have been sufficiently low so that ice would be condensed in the solids which would fall together to form the planetary body. Imagine an earthlike body with about four times as much mass in the form of water as of rock. Conditions would certainly be favorable for the origin of organic life, but the depth of the ocean would be some thousands of miles. It would be a prodigious feat for intelligent life forms on such a planet to erect a radio telescope at the surface of the ocean and engage in interstellar communication; such a feat would probably represent for them a technological effort comparable to our having put men on the moon.

Perhaps the main point to be made in this chapter is not only that the prevalence of planetary systems in the galaxy should be exceedingly high, but the great majority of these should have ages at least twice that of the solar system. Organic evolution and technological development should have occurred in such planetary systems a very long time ago, for the most part. If we do not find evidence of interstellar-communication networks having been formed by such extraterrestrial civilizations, then we should indeed be concerned about the prospective longevity of our species upon the earth.

Once again refer to Carl Sagan's formula for the number of civilizations in the galaxy, which is given in Ch. 1, and note the factor f_l, which is the fraction of suitable planets on which life originates and evolves to more complex forms. In the following chapter Cyril Ponnamperuma touches on this factor and shows that the chemical processes leading to the emergence of life are part of the evolutionary process in the universe. The transition from atom to small molecule to polymers of biological significance may be commonplace in the universe.

Over a hundred years ago vitalistic theories dominated the field of chemistry. It was considered heresy to subscribe to the idea that organic compounds could be generated in the absence of life. In 1828 Wohler demonstrated that urea, a product of animal metabolism, could be synthesized from inorganic species. Such experiments gradually eroded the barrier between the living and the non-living world.

Recent experiments on the synthesis of amino acids, purines, pyrimidines, and sugars have dramatically bolstered the Oparin-Haldane hypothesis that life is the result of the action of natural forces on the materials of the universe.

3

The Chemical Basis of Extraterrestrial Life[1]

Cyril Ponnamperuma   *University of Maryland*

Introduction

The problem of chemical evolution is central to any discussion of extraterrestrial life. In the whole sequence of events that could eventually culminate in our being able to communicate with an extraterrestrial intelligence, the question of the origin of life is paramount.

[1]Prior to publication, this seminar given in the summer of 1970 was updated, particularly with regard to the discussion of the Murchison and Murray meteorites, the Apollo lunar samples, and interstellar space.

To an exobiologist, the search for extraterrestrial life is the prime goal of space biology. Most emphatic is the statement from the Space Science Board of the U. S. National Academy of Sciences:

> It is not since Darwin and, before him, Copernicus, that science has had the opportunity for so great an impact on the understanding of man. The scientific question at stake in exobiology is the most exciting, challenging and profound issue not only of the century but of the whole naturalistic movement that has characterized the history of Western thought for over three hundred years. If there is life on Mars, and if we can demonstrate its independent origin, then we shall have a heartening answer to the question of improbability and uniqueness in the origin of life. Arising twice in a single planetary system, it must surely occur abundantly elsewhere in the staggering number of comparable planetary systems.[2]

If the discovery of microbes on Mars will be the fulfillment of our search for extraterrestrial life, hearing a single word from Alpha Centauri or Epsilon Eridani may be considered an achievement at the pinnacle of exobiology.

In the search for life beyond the earth, three approaches are possible. Within our solar system, planets could be inspected by remote or immediate sensors, or by man. Beyond our planetary system, contact with other civilizations would probably be limited to radio communication. For this approach to be successful, one must find civilizations with technologies comparable to our own. A third possibility is to study the problem experimentally in the laboratory. Using the earth as the example of the one laboratory in which life has occurred, we would like to extrapolate from events on the earth to events elsewhere in the universe. If the laws of physics and chemistry are universal laws, if the sequence of events that took place on the earth occurs elsewhere in the universe, we would then logically, perhaps, conclude that life exists elsewhere in the universe. Life could also have evolved to the point where intelligent beings may populate other planets around other stars.

Let us turn then to the basic question: how did life begin on earth? How did life begin in the universe? The question of the origin of life has until recently been a metaphysical preserve, but three factors have made the scientific study of this problem possible, both theoretically and experimentally: (1) astronomical considerations, (2) recent developments in biochemistry, and (3) the triumph of Darwinian evolution.

Implications of Recent Discoveries in Astronomy

As discussed by Dr. Cameron in the preceding chapter, there are many more opportunities for life to emerge in our universe than was previously

---

[2] "A Review of Space Research," National Academy of Sciences, National Research Council Publication 1079, 1962, Ch. 9.

believed. Estimates of the number of planetary systems suitable for life range from 10^8 to 10^{18}, depending on the number of restrictive conditions invoked. One is forced to conclude that there is nothing unique about our sun, which is the mainstay of life on this planet.

Recent Advances in Biochemistry

Although the phenomenon of life is manifested in forms as diverse as *E. coli* or an elephant, it is, basically, the result of the interaction of nucleic acids and proteins. Whether one considers the smallest microbe or the most intelligent human being, these two types of molecules determine the nature of life. Remarkably, only a few monomers constitute each of these important polymers. Nucleic acids are composed of five bases, two sugars, and a single phosphate group. Protein molecules generally consist of different combinations of only twenty amino acids. The alphabet of life is therefore extremely simple; the wide variety of life observed today may be traced to a mere handful of chemicals.

The Influence of Darwinian Evolution

The basic concept underlying our approach to the study of the origin of life is an evolutionary one. It is an outgrowth of the naturalistic movement which began during the Renaissance and peaked in the middle of the nineteenth century when Darwin put forward his theory of biological evolution. Darwin postulated a unity of the earth's entire biosphere, in that all living things are derived from a hypothetical "primordial germ" deemed to be the first life. More complex forms of life then evolved from simpler organisms over a very long period of time.

This revolutionary theory has now won general acceptance, and repercussions are still being felt even in fields of thought far removed from biology. In considering the early history of our earth from its formation 4.5 billion years ago to the time life first emerged, the physicist Tyndall wrote in 1871:

> He [Darwin] placed at the root of life a primordial germ, from which he conceived the amazing richness and variety of the life that now is upon the earth's surface might be deduced. If this hypothesis were true, it would not be final. The human imagination would infallibly look behind the term, and, however hopeless the attempt, would inquire into the history of its genesis . . . a desire immediately arises to connect the present life of our planet with the past. We wish to know something of our remotest ancestry . . . Does life belong to what we call matter, or is it an independent principle inserted into matter at some suitable epoch—say when the physical conditions became such as to permit the development of life?[3]

[3] J. Tyndall, *Fragments of Science for Unscientific People,* Longmans Green and Co., 1871.

In essence, the triumph of the theory of biological evolution has prompted the consideration of another form of evolution which preceded it, namely, chemical evolution.

Historical Considerations

When did life begin? Present evidence indicates that the earth is 4.5 billion years old, and, since the oldest evidence for life on earth has been dated at around 3.5 billion years, life presumably emerged sometime between these two dates. The age of the solar system is perhaps 5 billion years, and that of the universe is believed to be between 10 and 20 billion years. A cosmic evolutionary process can be postulated which spans the events from the genesis of the universe to the advent of intelligence. Three successive chemical stages have been suggested. During the birth of our star, inorganic chemistry would initially have predominated as hydrogen, carbon, nitrogen, oxygen, phosphorus, and other elements of the periodic table were being formed. When the solar system arose several billion years later, organic chemistry would have become more important. During this second stage, carbon compounds and small molecules such as amino acids, purines, and pyrimidines could have been synthesized and polymerized. As these macromolecules aggregated into precellular "eobionts," life could have emerged and, with it, the third stage, that of biological chemistry. The inorganic and organic phases may be considered chemical evolution.

The continuity inherent in the hypothesis of evolution implies a stage in which it would be very difficult to differentiate the living from the nonliving. As suggested by Pirie in his essay, "The Meaninglessness of the Terms Life and Living," it might be more useful if the terms "life" and "nonlife" were used in the same way chemists use the terms "acid" and "base." The concept of "hydrogen-ion concentration" allows the chemist to describe an acid as a solution which has a high concentration of hydrogen ions, whereas a base at the other end of the scale has a low hydrogen-ion concentration, but there is a continuity all the way through from acid to base. Perhaps a similar concept is needed to serve as a bridge from the nonliving products of chemical evolution to the living systems of biological evolution. Indeed, Lawden[4] has suggested that the presence of thinking protons and neutrons could perhaps be legitimately discussed if this line of reasoning is extended all the way back logically to the very beginning.

Spontaneous Generation

The idea of spontaneous generation, or the emergence of living organisms from inanimate matter, was widely accepted by the ancients. The philoso-

[4] D. F. Lawden, "Chemical Evolution and the Origin of Life", *Nature,* 202, 412, 1964.

pher Anaximander taught in the sixth century B.C. that the first animals arose from sea slime, and men from the bellies of fish. The ancient Hindu scriptures described life as having originated from nonliving matter. For example, the *Rig Veda* spoke of life beginning from the primary elements, while the *Atharva Veda* suggested the oceans as the cradle of all life.

Newton, Descartes, van Helmont, and others subscribed to this concept. Virgil's *Georgics* describes how a swarm of bees arose from the carcass of a calf. The Belgian physicist van Helmont provided a recipe for making mice:

> If a dirty undergarment is squeezed into the mouth of a vessel containing wheat within a few days, (say 21), a ferment drained from the garments and transformed by the smell of the grain, encrusts the wheat itself with its skin and turns it into mice. And what is more remarkable, the mice from corn and undergarments are neither weanlings nor sucklings nor premature, but they jump out fully formed.[5]

Such ideas could not withstand the advances of rigorous scientific thought. In the middle of the last century, Pasteur's experiments once and for all sounded the death knell to the theory of spontaneous generation at the morphological level, as it was understood by the ancients. Although Pasteur's experiments are often cited as an example of the triumph of reason over mysticism, it is curious that the theory of spontaneous generation has now been modified and refined to be used as the basis of contemporary studies in chemical evolution. On the molecular level, we now postulate the formation of molecules, proteins, and nucleic acids from an abiotic milieu.

Chemical Evolution—Early History

Charles Darwin summarized in a nutshell the whole concept of chemical evolution in a letter to his friend Hooker: "If we could conceive in some warm little pond with all sort of ammonia and phosphoric salts—light, heat, electricity, etc., present. . ., that a proteine compound was chemically formed ready to undergo still more complex changes. . ."[6] Subsequent laboratory experiments have essentially been nothing more than an attempt to substantiate this hypothesis. At the time Darwin wrote this letter, there was such a controversy over the whole question of the origin of the species, that little or no attention was paid to the problem of the origin of life. Besides, in 1861 Pasteur had proved that spontaneous generation was impossible; unfortunately, Pasteur's work gave rise to a widespread misconception that the problem of the origin of life could not

[5] J. B. Conant, (ed.), *Pasteur's and Tyndall's Study of Spontaneous Generation*, Harvard University Press, 1959.
[6] From G. deBeer, "Some Unpublished Letters of Charles Darwin," Notes and Records of the Royal Society, London, 1959, pp. 65–66.

be approached by scientific methods. The Dark Ages, so to speak, of chemical evolution followed, during which no respectable scientist was willing to do any experiments in this field. This trend was reversed in 1924 when the Russian biochemist A.I. Oparin reintroduced the concept of chemical evolution in scientifically defensible terms. He postulated the abiotic formation of a large number of organic compounds, their concentration by coacervation and other processes, leading to the ultimate genesis of life. In 1928, the British biologist Haldane, independently of Oparin, suggested the formation of a "hot dilute soup" by the action of ultraviolet light on the earth's primitive atmosphere giving rise to organic compounds soluble in the earth's oceans. The next great landmark is the essay of the British physicist Bernal, entitled, "The Physical Basis of Life," in which he described processes by which the organic matter formed by the Oparin-Haldane mechanism could have been concentrated in order to give rise to protein and nucleic acids.

The Primitive Atmosphere

A list of the raw materials available for chemical evolution can be seen in Table 3.1. The average composition of the solar system is reflected in that of the sun. Hydrogen predominates, and excluding the inert gas helium, oxygen, nitrogen, and carbon are the next most abundant elements. Ninety-nine percent of the biosphere is made up of hydrogen, nitrogen, carbon, and oxygen. In the presence of excess hydrogen, the equilibrium constants of these elements are such that carbon would be reduced to methane, nitrogen to ammonia, and oxygen to water. One can assume then, that on any juvenile planet, reducing conditions may have prevailed. This view is supported by the spectroscopic detection of methane, ammonia, water, and hydrogen in Jupiter's atmosphere. Jupiter is sufficiently large to have retained its primordial atmosphere.

Table 3.1   Composition of sun

Element	*Percent of Composition*
Hydrogen	87.0
Helium	12.9
Oxygen	0.025
Nitrogen	0.02
Carbon	0.01
Magnesium	0.003
Silicon	0.002
Iron	0.001
Sulfur	0.001
Others	0.038

The primitive atmosphere which prevailed during chemical evolution was probably the earth's secondary atmosphere. Whatever hypothesis we accept for the formation of the earth, whether it entailed a cold period of aggregation or a molten period, at some time the early primary atmosphere must have been lost, with the subsequent formation of a secondary atmosphere as the result of outgassing from the earth's interior. This secondary atmosphere was still reducing in nature. According to figures given by Holland,[7] it is reasonable to assume that 1 atm methane was present in the primitive atmosphere; this figure may have varied from 10^{-3} to 10 atm methane.

The conversion of the intensely reducing atmosphere present during the early stages of chemical evolution to our presently oxidizing atmosphere is represented in Fig. 3.1. The occurrence of free oxygen on earth is unique in our solar system. One explanation for its presence has been advanced. During the very early period when the reducing atmosphere prevailed, organic compounds were synthesized and ultimately washed into the oceans, which were thereby transformed into a primordial soup of organic compounds. The early organisms were probably heterotrophs which fed on the organic compounds thus formed. Due to the photodissociation of water in the upper atmosphere, an ozone layer was eventually formed. This ozone layer prevented the short-wavelength ultraviolet light from reaching the primordial oceans to serve as an energy source for further organic synthesis. As the food supply of organic compounds was gradually depleted and not replenished, a wholesale massacre of the early organisms occurred. Those organisms which were able to survive were presumably facultative autotrophs which had been fortunate enough to

Figure 3.1 Transition from reducing to oxidizing atmosphere.

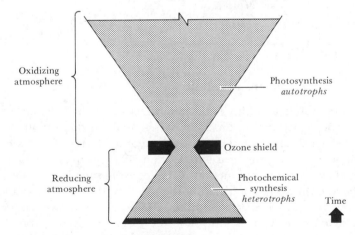

[7] H. D. Holland, *J. Geophys. Res.*, 66, 2356, 1961.

incorporate porphyrin-like molecules, enabling them to utilize the longer wavelength light which penetrated the ozone shield. Photosynthesis may have begun with these organisms, and from them developed the biosphere as we know it today.

Sources of Energy on the Primitive Earth

The kinds of energies available for the synthesis of organic compounds on the primitive earth are shown in Table 3.2. The sun provided the largest amount of energy, about 260,000 cal/cm²/yr. Ultraviolet light was the most important region of the spectrum at that time because of the necessity of dissociating methane, ammonia, and water. Although less energy was available at the shorter wavelengths, solar radiation quantitatively exceeded that provided by all the other forms of energy, such as electrical discharges, radioactivity, heat from volcanoes, and shock waves generated from meteorite impact. There were many different forms of energy available to act on the primitive atmosphere and give rise to those organic compounds necessary for the emergence of life.

Table 3.2 Energy sources available for organic synthesis on the primitive earth

Energy Source	Energy, $cal/cm^2/yr$
Radiation from the sun (all wavelengths)	260,000
Ultraviolet light, $\lambda < 2,500$ Å	570
$\lambda < 2,000$ Å	85
$\lambda < 1,500$ Å	3.5
Electric discharges	4
Cosmic rays	0.0015
Radioactivity (to 1.0 km depth)	2.8
Heat from volcanoes	0.13
Meteorite impact	0.1

Experiments in Chemical Evolution

Studies attempting to retrace in the laboratory the path of chemical evolution on the primordial earth have been based on the hypothesis that molecules which are important to contemporary forms of life were also important when life emerged. The two main objectives of the experimentors have been largely met. First, biologically significant small molecules have been formed under conditions believed to have prevailed on the primitive earth. Second, some of these molecules have been condensed, or polymerized, to demonstrate how the formation of macromolecules might have been initiated under the same conditions.

Synthesis of Monomers. Some of the first results in this field were obtained by Miller and Urey in 1953, when they exposed a mixture of methane, ammonia, water, and hydrogen to electrical discharges. Among the many organic compounds formed were glycine, alanine, β-alanine, aspartic acid, and glutamic acid. Much progress has indeed been made since the vitalism of 1828 when nobody was prepared to accept Wohler's synthesis of urea from inorganic sources.

Subsequent experiments using various mixtures of gases or simple precursor molecules, a wide range of energy sources (simulating those listed in Table 3.2), and reducing conditions have demonstrated the formation of more amino acids, fatty acids, monosaccharides, porphyrins, purines, and pyrimidines. Hydrogen cyanide is apparently a key intermediate in many of these prebiological syntheses.

Synthesis of Oligomers and Polymers. The formation of polypeptides and polynucleotides from their monomeric units is usually effected through a series of dehydration-condensation reactions. Condensation reactions which could have occurred under possible prebiotic conditions have been carried out both in the presence of water: to simulate the primordial ocean; and under anhydrous conditions: to simulate the dried-up bed of a lagoon.

Several dehydration-condensation reactions have been observed in the laboratory which could have occurred under aqueous conditions on the primordial earth. These include the photochemical synthesis of dipeptides and tripeptides; the formation of polymerized amino acids by the passage of an electric discharge through methane, ammonia, and water; the polymerization of some amino acids in an alkaline buffer during their conversion to activated adenylates; the formation of oligonucleotides in aqueous systems via a water-soluble carbodiimide; and the recently reported condensation of mononucleotides in aqueous solution at a neutral pH in the presence of cyanamide. In some cases, increased yields of oligomers were obtained when a template in the form of clay was added to the system. This observation confirms the suggestion made by Bernal several years ago that concentration and polymerization of organic molecules in the primitive oceans could have been facilitated by their adsorption onto inorganic clay surfaces along the seashores.

Under possible prebiotic anhydrous conditions, amino acid polymers and oligonucleotides have been produced in the laboratory. Fox has promulgated vulcanism (180-200°C) as a means by which amino acids could have been polymerized in the absence of water; the resulting amino-acid polymer is called "proteinoid" and is being extensively studied. The thermal phosphorylation (150°C) of nucleosides by inorganic phosphate salts has yielded dinucleotides and trinucleotides.

The phosphorylation and polyphosphate-mediated condensation of organic compounds on the primitive earth has recently been examined. One possible source of inorganic polyphosphates could have been derived from the thermal condensation of inorganic orthophosphates, which is readily accomplished at relatively low temperatures. In addition, it has been demonstrated that the 2'-, 3'-, and 5'-monophosphates are easily formed when an aqueous solution of adenosine is heated with linear polyphosphate salts.

The Interaction of Amino Acids and Nucleotides

The specificity of the interaction between monomers and polymers has been investigated in studies on the origin of the genetic code. Saxinger, Ponnamperuma, and Woese have used a chromatographic resin to which amino acids were covalently bound to demonstrate the relative affinities of nucleotides for several amino acids. A correlation was found between aromaticity and binding preference, whereas amino acids with only electrostatic binding potential did not display this preference. It was concluded that the major specificity of interaction in the polymer-monomer system is determined by monomer-monomer interactions.

Experiments Simulating the Planet Jupiter

Jupiter has been studied as an example of a planet where the evolutionary sequence that took place on earth in the distant past may be occurring now. Spectroscopically, the composition of the Jovian atmosphere resembles that which probably prevailed on the primitive earth. The calculations of Peebles and Gallet suggest that even though the clouds of Jupiter consist of frozen ammonia crystals, an inward temperature gradient is present due to the latent heat, making possible the presence of a water layer. Given the ingredients methane, ammonia, a possible layer of water, and a source of energy, it is reasonable that some kind of prebiological evolution may occur.

Laboratory experiments simulating the Jovian atmosphere have been conducted by passing an electric discharge through a mixture of methane and ammonia in a glass vessel; a cold finger, filled with liquid nitrogen and projecting into the interior, reduces the temperature to about 160°C. In addition to hydrogen cyanide, acetylene, nitriles, α-aminonitriles, and glycinonitriles are also formed; hydrolysis of the C-methyl and N-methyl derivatives of the glycinonitrites can yield amino acids. The red organic polymer generated in the experiments may explain the red colors observed in the planet Jupiter. This experiment demonstrates that, under Jovian conditions, some of the organic compounds necessary for life may be formed; the kind of prebiological evolution presently discussed as having taken place on earth four billion years ago may be currently taking place on the planet Jupiter.

Search for Molecules of Biological Significance in Ancient Rocks and Sediments

In addition to the synthetic experiments just described, another approach has been to study chemical evolution by examining sediments, meteorites, interstellar space, and lunar samples for evidence of life or the organic compounds which may have preceded its genesis. Figure 3.2 is a picture of the geologic clock which once again emphasizes that the early stages of evolution up to the beginning of the Cambrian period lasted a relatively long period of time. Man appears only during the last few minutes of this representation. But, for the student of the origin of life, the area of greatest interest occurred prior to that time. Fortunately, there are various landmarks. Samples from the Bitter Springs in Australia, dated at around 1 billion years ago, the Gunflint chert in Ontario, approximately two billion years old, and the Fig Tree and Onverwacht formations in South Africa, which are about 3.1 to 3.5 billion years old, have been

Figure 3.2 A geologic clock showing the stages in evolution.

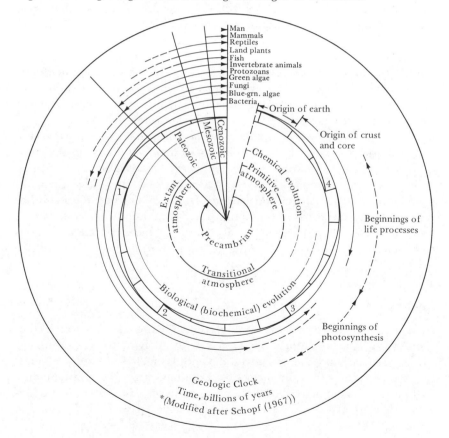

examined. Molecular fossils, rather than fossilized organisms, have been sought, which could retrace how life emerged. Even in the oldest rocks examined, dated at about 3.5 billion years, evidence for life has been found. Some of the microstructures formed in the Bitter Springs from Australia have been characterized by Barghoorn and Schopf as resembling blue-green algae. Other microstructures have also been found in the Gunflint chert and in the Fig Tree formation and the Onverwacht sediments dated at 3.4 billion years.

Through painstaking efforts, the chemical nature of some of the molecular fossils present in these rocks has been determined. In the 3.1 billion-year-old rock from the Fig Tree formation in South Africa, many biologically significant molecules have been identified, including normal alkanes, isoprenoid hydrocarbons, porphyrins, and amino acids. In addition, the ^{13}C to ^{12}C fractionation suggests biological activity. It must be remembered that our evidence is only circumstantial, for although the rocks themselves are 3.1 billion years old, there is no absolute method of establishing that the organic molecules contained therein are 3.1 billion years old. Although pieces of rock were chosen for study which were apparently impermeable and seemed to be well defined, the possibility that these organic compounds are contaminants derived from relatively recent microorganisms has not yet been definitely excluded.

Going back one stage further, one can look not simply at the rocks but at the organisms themselves. There may be organisms which were presumably very low in the evolutionary tree whose descendants may be existing today in a relatively unchanged form. The composition of these organisms may provide some insight into how life was formed or evolved. An example is the living microorganism *Kakabekia barghoorniana,* which is morphologically very similar to a microfossil found in the Gunflint chert in south Ontario. The contemporary *Kakabekia* was first cultured from a sample of topsoil in West Wales in 1964, and since then has also been found in Alaska and Iceland. In the laboratory, *Kakabekia* has been cultured in saturated ammonia-water containing nutrients such as glucose; it does not require oxygen. Remarkably, these cultural conditions resemble the environment believed to have prevailed on the primitive earth.

Meteorites

Although meteorites have been analyzed for the presence of organic compounds for over a century, the source of the polymeric organic matter detected in carbonaceous chondrites has not been clear. In previous studies, the possibility that the meteorites examined had been contaminated with terrestrial biomolecules before being analyzed could never be excluded with certainty. However, amino acids believed to be indigenous to the Murchison and Murray meteorites, which are both type II carbonaceous chondrites, have recently been detected. When the analytical techniques of ion-exchange chromatography, gas chromatography, and gas chromatography combined with mass spectrometry are used, the presence of six biological amino acids (glycine, alanine, valine, proline, glutamic acid, and aspartic acid) and eight amino acids normally absent in protein (N-methylglycine, β-alanine, 2-methylalanine, α-amino-n-butyric acid, β-amino-n-butyric acid, γ-amino-n-butyric acid, isovaline, and pipecolic acid) have been detected in both meteorites. In addition, three amino acids (N-methylalanine, N-ethylglycine, and norvaline) which were only tentatively identified in the Murchison, were positively identified in the Murray meteorite. [8]

Several lines of evidence support an extraterrestrial origin for these amino acids. The D and L isomers of those amino acids having asymmetric carbon atoms are almost equally abundant, whereas terrestrial biological contamination would have resulted in a predominance of those amino acids commonly found in protein. A random abiotic synthesis is indicated by the presence of several nonprotein amino acids, and the identification of all the isomers of the amino acids composed of two and three carbon atoms, and all but two of the aliphatic isomers with four carbon atoms.

The aliphatic hydrocarbons of the Murchison meteorite have been found to consist largely of saturated alkanes. Their gas chromatographic traces resemble closely those synthesized by the action of electrical discharges on methane, and mass spectral data have revealed the same dominant homologous series in both samples. Isotope-fractionation studies indicated that, except for the carbonate, the Murchison meteorite is enriched with ^{13}C in every fraction. All in all, these results indicate that the amino acids and aliphatic hydrocarbons detected in the Murchison meteorite are indeed indigenous.

Other organic compounds found in the Murchison meteorite have also been examined. The extracted aromatic hydrocarbons resemble those produced by the pyrolysis of methane, in that both consist largely of polynuclear compounds containing an even number of carbon atoms. A search for purines, pyrimidines, and triazines yielded only 4-hydroxypyrimidine and two of its methyl isomers. The absence of the more common biological heterocyclic compounds supports the contention that the organic matter found in the Murchison meteorite is not the result of biological contamination.

Taken together, the results obtained in studies of the Murchison and Murray meteorites strongly suggest that extraterrestrial chemical evolution has taken place—and may be continuing. This conclusion is reinforced by the recent discovery of many organic molecules in interstellar space.

[8] Many more amino acids have been identified in subsequent studies.

Interstellar Matter

Observations at the National Radioastronomy Observatory in Green Bank, West Virginia during the last three years have revealed the presence of several organic molecules in the intergalactic regions. To date, the molecules discovered include H_2O, NH_3, HCHO, CO, CN, H_2, HC_3N, CH_3OH, CHOOH, CS, NH_2CHO, SiO, OCs, CH_3CN, HNCO, HNC, and CH_3C_2H. Polyatomic molecules such as these are now considered to be widely, but very thinly, distributed throughout interstellar space. The nature of the complex molecules observed indicates that they are all steps in a widely occurring evolutionary sequence which spans the formation of atoms during the birth of a star to the synthesis of biologically significant molecules making up the early atmospheres of the planets.

Lunar Samples

The Apollo 11 and 12 lunar samples taken from the Mare Tranquillitatis and the Oceanus Procellarum contain only 200 ppm carbon. Although there is no conclusive evidence that any hydrocarbons, fatty acids, amino acids, sugars, or nucleic acid bases are indigenous to the lunar surface, samples from other sites and/or below the lunar surface might yield different results; this possibility will be explored in future Apollo samples.[9]

Conclusion

In summary, a composite picture of the origin of life is being formed, both from synthetic experiments and from analytical studies. Evidence is accumulating for the hypothesis of chemical evolution, generally known as the Oparin-Haldane hypothesis. One could argue then that if conditions similar to those on the primitive earth prevail elsewhere in the universe, the same kind of events might happen, resulting perhaps in a kind of life very similar to ours. Cyril Darlington, the noted British biologist, once said that if someday we land somewhere else in the universe, we should not be surprised if we are greeted by beings 5½ ft tall with two eyes in their forehead because there is such an advantage in that posture.

[9] Analyses of Apollo 14, 15, 16, and 17 samples have yielded similar results.

In the formula for the number of civilizations in the galaxy presented by Sagan in Ch. 1, there appeared the factor f_i, which represented the fraction of life-bearing planets with intelligence possessed of manipulative capabilities. The following chapter, by Michael Arbib, discusses a part of the problem associated with the evaluation of this factor. He is concerned with the stages of evolution upon the earth which lead to the higher animals and culminate with intelligence in man.

However, the term "intelligence" is a complex one which raises many questions. What does one mean by it? Can one be sure that one understands a natural phenomenon before the features of that phenomenon have been duplicated in the laboratory? One must also be concerned with the propagation of intelligence. When might a machine be considered intelligent? If a machine can be programmed to reproduce itself out of a carefully structured environment, it is probably not very intelligent, but if it is programmed to reproduce itself out of a wide variety of environments which are encountered in nature, then it must certainly have most of the attributes that we attribute to intelligence in human beings. Surely man is more than just such a machine. However, it is doubtful that man can fully understand himself until he is able to produce such machines. These are some of the questions raised by Arbib in his discussion of the evolution of intelligence in the following chapter.

4

The Likelihood of the Evolution of Communicating Intelligences on Other Planets

Michael A. Arbib   *University of Massachusetts at Amherst*

The Story so Far

The series of papers compiled in this text focuses on three questions: First, are there "intelligences" elsewhere in the universe? Second, if there are, can we, at least in principle, communicate with them? Third, if such communication is possible, how might we implement it? In the first paper, Carl Sagan proposes that our search for intelligence focus on the possibility of a planet-based life made up of large complex molecules. Cameron, in the second paper, suggests that most stars probably do have planets, and Ponnamperuma, in the third paper, states that it is almost inevitable that if the orbit of a primitive planet lies in a suitable temperature range, then organic chemicals, "the building blocks of life," will form. When we convince ourselves that intelligences can evolve on other planets, we will have found a lower bound to the evolution of intelligence in the universe, because strange life forms may even drift among the stars—as those who have read Hoyle's *Black Cloud* may well imagine.

We shall also be interested in determining whether an intelligence that evolves is likely to create a technology, because it is probable that technology is required for interstellar communication. Although we evolved biologically to communicate using sound waves, it is conceivable that another species could have evolved means of electromagnetic communication.

Our task in this chapter is to understand to what extent one might expect intelligence to evolve, and to that end we must consider what we mean by "intelligence" and what we mean by "evolution." Some years ago a book appeared with the title, *Is There Intelligent Life on Earth?*. We shall assume that the answer is yes and use the evolution of the human brain as a paradigm for the way intelligence might have evolved. However, it must be stressed that much of the discussion will be in a realm far removed from the careful results described in the two previous chapters. It is one thing to simulate the motion of thousands of particles on a computer and see that they form into a disk which breaks up into a configuration akin to a star and several planets, or to pass ultraviolet light through a nearly empty jar and see that after a while some simple organic chemicals form. It is a vastly different thing—and inconceivable with present technology—to do the equivalent of taking a jar and energizing it for a few billion years to see if eventually intelligent life will be produced. At best, then, this chapter will attempt to make it plausible that, given that few billion years, it is almost inevitable that some sort of intelligence will evolve on any planet on which autocatalytic reactions have arisen. But whether that intelligence is anything at all like that of humans will be a very open question, and our attempts at an answer will at times be closer to science fiction than to science.

Evolution

The notion that life *evolved* on earth rests on the hypothesis that our planet initially had little in the way of chemical complexity, and by gradual changes, all of which are completely explicable by physical law, the complex organisms that we find on earth today have arisen.

There are two basic ingredients in the current theory of the evolution of life on earth. Darwin gave us the concept of natural selection—that organisms could be "selected" *naturally* for survival. For example, animals that were more fit in some way to live in certain environments would tend to reproduce more than other animals in that environment so that over the passage of time the animals that existed would be more and more adapted to the "ecological niche." The *ecological niche* is the complex of circumstances which provide the life space for a particular species. We should note that the ecological niche is not specified simply by temperature and climate, but also by what other organisms, plant or animal, live in the area, because as soon as a new type of organism evolves, there is then the possibility of evolving yet another species to live in the interstices and to exploit the new living relations thus made possible. Thus we have a proliferation of complexity of organisms with time.

The other ingredient in the theory of biological evolution is Mendel's idea of the *gene*. In modern terminology we would speak of the genes as forming the program which directed the growth of the organism from a single cell in interaction with the environment. If an animal lived long enough to reproduce, then portions of its genes would provide the program for the growth of its descendants. Note that this says that it is not any peculiarities the animal may have acquired in its lifetime that survive, but rather the "program" which survives. If an animal had acquired a long neck by stretching to nibble at trees, it could not pass that on to its child. However, if a change in the program (a mutation in its genes) gave it the character of being more likely to grow a long neck in a certain range of environments, then one could expect the offspring to also have a long neck. We thus have the distinction between the *genotype*—the type of growth program—and the *phenotype*—the type of the actual grown organism, which results from both the genotype and the environment in which it operates. The phenotype determines whether or not the creature will live to reproduction age and pass its genes on, but the actual changes from generation to generation are within the genes.

In the preceding chapter we are reminded that life on earth is built up from very simple building blocks. DNA, which we now believe to be the material from which genes are made, has only four or five "letters" in which the "programs" are written to direct the growth of organisms, while the actual building from these programs is done with amino acids, only twenty in number. It was suggested from this evidence that life came from a single precursor, but we must be a little wary in accepting this assumption, since the fact that only one type of basis "won" out does not mean that there weren't other "candidates" initially. Thus life forms on planets of other stars might be quite different even in this basic structure, though there may be chemical reasons to expect DNA to be ubiquitous. It might also be asked whether it is necessary for *any* (nonterrestrial) life form to have a genotype distinct from a phenotype, in other words, whether we have to have a program to direct growth and change, or whether in fact the organism might be able to reproduce itself as a whole. One might imagine some planet in a distant galaxy whose beings reproduce by xerography with no gene required! For some mildly convincing arguments in favor of the gene concept, see the section entitled, "What is the Role of Descriptions?" in Arbib [1969a]. When we try to extrapolate from the careful science of what exists on the earth, we get into the realm of science fiction so that the speculations in a book like Jose Farmer's *Strange Relations,* which explores some of the more bizarre forms that symbiosis and sexuality might take on other planets, may be as right—or as wrong—as those of the scientist.

Why Not Be a Rock?

At this stage, we should ponder the crucial question that evolutionists so rarely ask. It is so common to fantasize that evolution is a process existing for the sole purpose of producing human beings—that amoebae have been struggling out of the slime for millions of years trying to aggregate into human brains—that people never ask the naive question, "Why not be a rock?" If sheer survival and long life is the goal, what is superior to a rock? A rock has no problems, and even if, at the end of billions of years, a convict comes along with a sledge hammer and smashes the rock, at least it has a billion years of just sitting, which far transcends the human three score years and ten! So why aren't we all rocks? Unfortunately, earthly chemicals took a "wrong turn" a few billion years ago, along the lines suggested in the preceding chapter. Once some moderately complex chemicals form, it is likely that they will form aggregates that are autocatalytic: reactions occur in which a chemical triggers the production of more chemicals like itself. Once that epochal "mistake" was made, there was no turning back. It was not that all these chemicals were striving to be Man; there was no mystical goal or final cause which was "pushing them on." Rather, once there existed complex chemicals which were able to produce chemicals like themselves, there was always a probability that the resultant chemicals would be changed somewhat. There was then a small chance that a compound would arise that could produce more copies of itself than the older ones could. The whole of evolution was at that stage unleashed, and it is interesting to see to what extent life as we know it must then result.

Life as we know it is based on cells. This seems to make sense because a membrane surrounding chemicals in reaction can exclude some chemicals, while containing precious enzymes that can speed up the reactions. Again, a plausible case might be made for the emergence of DNA or some sort of general message to direct things. Hence there are plausible reasons for suspecting that, in the long run, any planetary form of life starting from autocatalytic reactions would eventually come to be—at least crudely—a multicellular form such as we know on earth. However, before we trace the development of a series of earth organisms which will give us some idea of the complexities that cells can enter into in their relations with each other as they move from single-celled animals to the intelligence of man, let us briefly talk about self-reproducing machines.

Self-reproducing Machines

There are two reasons for briefly discussing self-reproducing machines here: one is that they will give us some insight into the processes of reproduction and evolution as physical processes; the other is that self-reproducing machines may well play an important role in interstellar communication.

As students of Southeast Asian politics know, it is widely believed that it is very simple to produce a self-reproducing machine of the type shown in Fig. 4.1. If the environment does indeed consist of a string of dominoes, and the system to be reproduced is a falling domino, reproduction can clearly take place. However, from our point of view this model is only at the level of simple chemical reactions that, once started, can continue to maintain themselves. What we are really interested in is whether really complicated systems can reproduce themselves in a way that we can understand on the basis of our physics and mathematics without introducing some mysterious distinction between living and nonliving forms.

Von Neumann [1951] noted that when we look at biological reproduction, the offspring that is produced is at least as complicated as the parent, but when we think about production lines, with machines producing other machines, we tend to expect a degradation of complexity. It is thus of great interest that Von Neumann was able to show that one could, in fact, have machines of arbitrary complexity which would reproduce themselves. We shall not go into the details here, but we can at least examine two figures which give an idea of the strategy, while omitting the detailed program.

With much careful programming one can specify the "universal constructor" A of Fig. 4.2, which when given a program I_N for constructing any other machine, N, will read through that program, compute upon it, piece together components from its environment, and construct a copy of the machine N, as shown in the right-hand side of Fig. 4.2.

Figure 4.1 A simple self-reproducing machine.

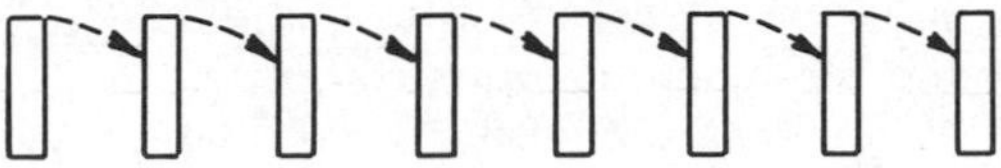

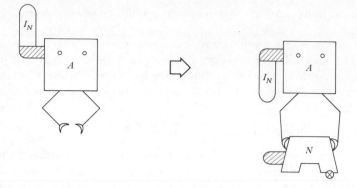

Figure 4.2   The "universal constructor" A; when given a program I_N for constructing any other machine N, A will read through that program, compute upon it, piece together components from its environment, and construct a copy of the machine N, as shown in the right half of this figure.

The procedure is somewhat unbiological—it is as if a mother, instead of starting with a single fertilized cell and growing it in her womb for nine months, were instead to start by piecing together the toenail and give birth to the child as the last hair was put in place on its head. There is now active research into how we might make "biological perturbations" of this theory, but here let us ask how we might modify A to obtain a machine capable of reproducing itself. One might think that the strategy would be to give A a description of A. Then A would compute upon I_A to yield as the N of Fig. 4.2. a copy of A itself. However, A equipped with a program has reproduced A without a program, so that it has produced a "jackass" as offspring rather than reproducing the initial configuration. The resolution comes from noting that in biological systems, a cell, before reproducing itself, will copy the genetic message, so that when it splits in two, one copy can go to each daughter cell.

These considerations yield the self-reproducing automaton of Fig. 4.3, in which A is augmented by two new subsystems B and C. When activity is initiated, the first subsystem to work is B, which makes a copy of the program I_D and inserts it in the holder at the top of B and C (see Fig. 4.3(a) and (b)). Then B transfers control to our universal constructor A which works through the program I_D. If we choose the described machine D to be in fact the machine $A + B + C$ without programs, then A builds the connection of A, B, and C (Fig. 4.3(b) and (c)). Having completed its construction, A transfers control to C which now takes the spare copy of the program I_D and transfers it to the "offspring" (Fig. 4.3(c) and (d)) and self-reproduction is complete.

Rather than go into details of this theory (for which, see Von Neumann

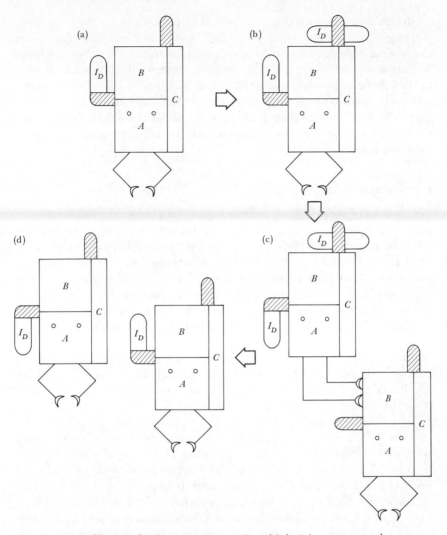

Figure 4.3 Self-reproducing automaton, in which *A* is augmented by two new subsystems *B* and *C*.

[1966] or Arbib [1969], Ch. 10), let us devote the rest of this section to some speculation about the role self-reproducing machines might play in interstellar communication. Developments in artificial intelligence (for example, as discussed by McCarthy in Ch. 5) raise the question of the extent to which our eventual interstellar communication will be directly with living creatures, and the extent to which it will be with man-machine symbioses or even with a purely machine intelligence alone. In any case, much of the discussion of interstellar communication posits radio communication as the basic medium. However, we might well imagine trying to design a self-reproducing machine that carries out its own synthesis starting from the interstellar gas. These machines could then reproduce every time they travel a constant distance out from the home planet to yield a sphere moving out from the home planet with a constant density of these self-reproducing machines. Before leaving this line of speculation, we might ask: What if they start to mutate? Will it matter? Are we so lonely here on the planet Earth that we want to talk to anybody from out there, or must we be assured that we are getting accurate messages from another culture?

Intelligence

We now have some idea of the mechanism of evolution, which involves creatures reproducing and mutations (and other effects) changing the essential program for their growth, so that over time, certain types of programs become more common than other programs in a given ecological niche. What then is the "intelligence" that overly romantic writers have viewed as the goal of earthly evolution? For one thing, intelligence is not a single thing, and the present controversy on IQ tests reminds us that intelligence has many dimensions. It may well be, then, that on planets of other suns, many of the dimensions of intelligence will be different from those we know. However, we can isolate some essential characteristics of intelligence which may provide the substrate on which specific talents like musical intelligence, mathematical intelligence, plain horse sense, and so on can be erected.

To understand the characteristics of intelligence, we should consider that the processes of evolution yield creatures who in some sense model their world better and better, or at least those aspects of the world relevant to survival in some ecological niche. An organism is impinged upon by the world around it and carries out actions upon its world. It survives to the extent that its actions are appropriate to the world in which it lives. If an amoeba were to always extend itself into acid and away from food, there would not be any amoebae left. Thus to some extent even an amoeba can be viewed as a representation or "model" of the fact that acid is dangerous, and certain chemicals correspond to food. As we come to more and more sophisticated organisms (see the following section) we find more and more aspects of the environment are "captured" in the repertoire of behavior of the organism.

In this vein we might say that what characterized human intelligence is that instead of having a single "model of the world" wired in genetically so that all we can do is live in a limited environment and react in a stereotyped way, we are able to learn as we grow, to find out more and more about our world, to adapt ourselves to differing environments, and to learn skills that we can transfer to situations where we have never used them before. It appears, then, that the evolution of intelligence is a natural consequence of the evolution of living forms—as organisms evolve that are better adapted to react to the flux of the environment about them, so may we expect the occurrence of organisms able to recall things about their environment and to make plans and strategies on the basis of which they can act.

It is thus not implausible (though this discussion has done scant justice to all the subtleties) that once evolution "gets going" in biological systems, the evolution of intelligent organisms is pretty much inevitable, given enough time. Intelligence involves the ability to hypothesize, to try different things out, and to choose between alternative futures.

One of the interesting facts about the earth, which gives us some hope for thinking there may be more in common between intelligences in the planets of different stars than a quick browse through science fiction in its less anthropomorphic forms might suggest, is that on the earth visual systems have evolved independently in quite different species (see Gregory [1966]). There exist two general types of eye, the compound eye of the insect and the simple eye, with its single lens focusing light upon a whole array of cells, that we see in man. The "simple" eye occurs also in the squid and the octopus, even though they have evolved in completely different ways from man. Thus the fact that completely different paths of evolution—though admittedly they all occur on the planet Earth—have led to the evolution of similar structures suggests that there may well be many commonalities between creatures who have evolved to live on a planet illuminated by electromagnetic radiation—especially among creatures who must move around when subject to gravity. As we come to understand evolutionary theory better, we may actually be able to predict such commonalities. However, the gross similarities of visual systems should not blind us to crucial differences of detail (as we shall see below), and the diverse forms among terrestrial insects alone should remind us that vast differences will remain which can seriously hinder types of communication.

From Amoeba to Man[1]

Let us now give a short overview of the evolution of human intelligence. To start with, we indicate in Fig. 4.4 the basic organization of the computing system which underlies all multicellular organisms. Receptors take light, sound, and touch energy from the environment and convert it into electrical pulses which can propagate down various signalling lines to impinge upon various neurons. These neurons have already been changed

---

[1] This section has drawn heavily upon the works of Buchsbaum [1948] and Elliott [1969]. The reader is warmly recommended to turn to these two excellently written volumes to supplement our somewhat sketchy account. See also Altman [1966] and Herrick [1926].

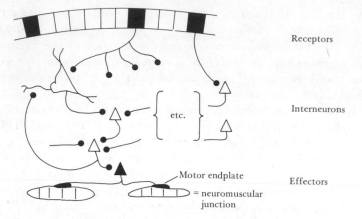

Figure 4.4   The basic organization of the computing system which underlies all multicellular organisms.

by previous activity, and so they react on the basis both of what is coming in and the current internal state of the system. The activity of various neurons inside the net will then impinge upon muscles or glands, and so change the way in which the organism acts. We shall study forms of life that exist on the earth today to get some insight into how this sort of computation structure, which is at the root of intelligence as we know it, could evolve.

Our brains contain approximately 100 billion neurons, in addition to all the other cells that make up our bodies, so that our intelligence results from the orchestration of the interrelated activity of billions upon billions upon billions of single-celled creatures. Yet the single-celled amoeba lives today and is a very successful animal that can do very well as long as it finds food to live on. An amoeba can respond to the chemical gradients set up by a piece of food by extending pseudopods to surround the piece of food, closing upon it, and digesting it. It has the basic reactivity to avoid harmful chemicals and move towards chemicals that signal food. In its own ecological niche it does as well as man in his. (Man has a far broader and more varied ecological niche, and that is where intelligence enters.)

However, the amoeba is limited in that if bits of food were near two parts of the amoeba at the same time, it would try to extend toward both of them. The obvious "engineering improvement" is to provide the organism with the ability to coordinate its different parts.

A paramecium, which is still only a single-celled organism (remember that humans contain billions of cells) is already rather sophisticated. It is surrounded by little hairs called cilia. Whereas in humans each hair comes from a separate cell of its own, cilia are outgrowths of the single cell. These cilia can beat together to move the paramecium about, and in fact there exist microscopic filaments that connect the roots of the various cilia to yield coordinated movement. If the filaments are cut, the cilia wave in an even sillier fashion. Again, while the amoeba will ingest food wherever it makes contact, the paramecium has a distinct oral groove lined with cilia which can beat together to propel bits of food down the groove. Here we have a *highly evolved* creature which has been evolving as long as we have to live in a simple environment in which, because of its coordinated action, it can make a somewhat better living than amoebae do. In fact, a paramecium exhibits a rudimentary memory in that if it bumps into an obstacle it can compensate: it backs off, turns a little bit, and goes forward again, and if it bumps into something, it repeats the routine, it backs, turns a little bit, and goes forward again. Thus it has a little bit of memory: it can remember whether something it did failed, and do something else. Of course, it isn't sophisticated enough to know that if it has gone round through 360°, it has had it; it will just keep turning round and round until it dies.

Although we have been discussing creatures that exist today, we are perhaps recapitulating some of the development that must have occurred in that branch of evolution which led to man. In the amoeba, we see the basic ability to respond to the crudest stimuli, both nutritious and noxious. The paramecium further exhibits coordination—the organism can begin to act as a whole—and has a rudimentary memory in that what it does now depends somewhat on what it has just done. However, this is a far cry from the human ability to build on a lifetime of experience. To gain more insight we must turn to the sponge, the simplest of all living multicellular creatures.

With hindsight, we can see good reasons for the evolution of cellular organisms. Of course, the first colonies of cells did not know that they were forming multicellular organisms; it just happened that such groupings tended to survive because at the time they were able to exploit the environment a little bit better than their contemporary single-celled organisms could do. Sponges comprise a colony of cells, each somewhat like little paramecia, but which are embedded together in a single matrix, where they cooperate to the extent that their cilia bear together to force a continuous stream of water that goes in through little incurrent pores and out through a large excurrent opening. The various cells, instead of moving individually at random to occasionally encounter food, together create a current that is likely to bring food to them all. However, despite this cooperation to "encourage" food to move in their direction, there is little real coordination; there is no nervous system that links all the different parts together. If something noxious bumps into one of the pores, then the cells around the pore can contract to stop that noxious thing from getting inside the sponge, but there are no paths to enable related action elsewhere in the organism—the sponge is like a multicellular amoeba, in that it can only react locally.

The hydra is the simplest multicellular organism with "global coordination." It lives in fresh water ponds with its base moored on a rock, and it has little tentacles with which it can grab little things as they come by and stuff them into its gut. However, the important point is that it has a nervous system, albeit a diffuse and rudimentary one, which allows the effects of a stimulus to pass through the whole organism so that the whole organism can bend away from a noxious stimulus. In other words, the whole organism can begin to act in a coordinated way. However, there is little detailed computation; if you "hit" it hard enough, more of it will get out of the way, but there is no wherewithal to decide what is the appropriate action, or how to pool different stimuli.

In the flatworm we at last find a nervous system which does more than just carry a shock further and further away from the point of stimulation; it can also combine different types of stimuli. The flatworm does not have pattern-sensitive eyes, but rather photo-sensitive cups, which these animals can use to detect which side is brighter, so that it can move away from light by signaling a contraction to the muscles on the side of the body away from the light. Its nervous system has a great deal of structure to it and can carry signals in an orderly manner down the two sides of the animal. We also see the beginnings of a brain—the flatworm has a head end with a greater amount of computation, not very much, but just sufficient to take account not only of information about chemicals and touch, but to pool it with the primitive visual information that comes in from the photo-sensitive spots. The nervous system is sufficiently sophisticated that if the animal is moving along a stream and feels the pressure from the stones underneath, smells rotting meat, and is exposed to a light from the same side, it can pool all the effects to compute which way it will turn. Here we see all the ingredients of the real nervous system as we know it.

We close this section by briefly noting the two major differences between the nervous systems of flatworm and man. The first resulted from the evolution of more and more sophisticated types of *vertebrate* organization, so that we have a single spinal system which can relay information from all over the body to the head end where it can be pooled and processed by the brain. Thus the organism becomes a coordinated system which acts much more as a whole than the invertebrates which comprise, to some extent, a collection of subbrains, each controlling a separate portion of the animal with relatively slight constraints as to how it acts overall. The second resulted from the evolution of the *mammals*, where the keeping of the embryo in a womb where metabolic conditions could be kept more constant until it was born was coupled with the evolution of a new out-growth of brain called neocortex, which gets bigger and bigger until in man the huge outfolding of neocortex covers all the "older" brain structures. It is the neocortex which seems to contribute most to our ability to remember as we do, to plan, and to use speech.

Cultural Evolution

With language, man acquired the substrate for culture, for it allowed the fruits of intelligence to be pooled and shared with the other members of the species without using the slow genetic route. Today, cultural evolution seems to dominate the biological processes discussed previously. Men —or at least man-like species, the hominids—have changed immensely in the last million years. They evolved the upright gait which has freed their hands for the delicate manipulation which has allowed them to exploit tools; since the hands are free, it is unnecessary for them to fight with their mouths as many animals do, and the mouth has been freed for the subsidiary evolution of the vocal apparatus.

It is hard to predict whether man's biological evolution will continue. We can make suggestions about what sort of social organizations may be needed to stop the crises that come from war, overpopulation, and so on, involving greater levels of social awareness and coordination. It would be pleasant to think that there might be some biological mutations that would make man a more genuinely social creature and remove certain aspects such as aggression, which may have been useful for primitive hunting tribes but are not for modern society. But we can't predict that this will actually happen. We have stressed that evolution is built on local adaptations for better exploitation of an ecological niche, and must not be regarded as a giant progression whose goal is man. In particular, there is no guarantee that any future biological evolution would bring us closer to a more enlightened society. However, though biological evolution has no single goal, we can impose goals, and the hope is that we are now conscious enough that we don't have to be bound by just the chance interactions of different species in an ecological niche, but can begin to decide what things we would like to be important in order that we can create goals, and then we can begin to structure our society to achieve them.

In the historical development of man, we see less of the biological operation of natural selection changing physical structure and more and more of the building up of symbol systems that can be shared. No longer are the "messages" encoded in the genes to be passed on only from parents to child. Today our evolution seems to center more on the evolution of scientific systems, systems of different professions, systems of entertainment, systems of culture and religion, and so on, that can be passed from person to person by such mechanisms of nongenetic transmission as education. Again, technology has enabled us to increase the range of environments in which we can survive. This widening of the ecological niche is one of the most notable achievements of mankind. But while man is physically similar in many niches, we see dramatic differences in culture. Bees also have a culture, a social system, where each organism has a part in an overall scheme, but human culture is different in that having language, we have some overview of our culture. Even if we cannot understand everything about society, we can at least try to make some sort of judgment about it and try to describe to others at least the outline of what is happening, even if we cannot master many details. Might we, then, be able to communicate something about our way of life, our society, and our science, though, to the creatures of other planets? To succeed, they must have evolved to the stage in which their model of their world (see the preceding discussion) is based upon nongenetic transmission, for if the only way they can receive information from other creatures is by being born as their offspring, we have little chance of talking to the people of Betelgeuse!

In teaching someone to drive, linguistic instructions let you put a person into a car, and as long as you are in a fairly quiet street, you are able to give him the necessary instructions. On the other hand, you are not going to let him go into peak-hour traffic until he has added to that knowledge a great deal of refined motor coordination that involved brain structures of much older evolutionary origin. Language can get the person into the right "ballpark," while older systems can refine these activities to a level of skill. Together these play a crucial role. If you were driving across a piazza in Naples, you could not speak sufficiently fast to encode in a normal sentence the fact that there was a car coming at 45°, another car coming at 33 mph at 21°, and so on. You have to rely on older preverbal visual mechanisms that evolved beautifully to coordinate spatial patterns of sensory information with spatial patterns of action. Language evolved on "top" of that to get the right sort of "ballpark" of operation which gives people the basic idea of driving. Then they can use their "older apparatus" to refine the parameters and use special information efficiently.

Often, when we talk about language, we think so much of language as *the* human ability that we forget that the reason it works so well is that it exists on top of a substrate we share with other humans, which is not very easy to verbalize. Acknowledging this substrate emphasizes the problem of talking to creatures in a language that is universal, in the sense that if we give them sample programs, they will be able to compile our language into their own "machine language." Eventually, we may be able to communicate with them at some level, but the problem is to find how to express things in a "machine language" so simple that any technological civilization must be able to comprehend it.

What Can Be Communicated?

In terms of communicating with a far distant civilization, let us note that by "human intelligence" we do not mean the intelligence of any one particular human; what man knows is not what any human knows. When we communicate with a race on another planet, we will not be communicating with any one individual, but rather will be communicating with the whole species. In some sense because our communication is a message in linguistic form, it will presumably be something that a single individual could *comprehend* and could follow through—even if it was a 50-yr-long message. But the *creation* of the message will require a distillation which no single individual could devise. The difference becomes clear when we note that it is much easier to read a book than write a book—and there are many, many subjects in which one could read and comprehend a well-written book, but this is quite different from being able to write a comprehensible book in that area. Given that the message will in some sense be the distillation of the species' knowledge, it will presumably involve a huge amount of correlating techniques, assessment of intelligibility (which could require computer-run surveys), and so forth. It may well take so much technological input in addition to the contribution of individual experiences to complete it, that no individual will fully comprehend why that particular message is the best we can send, though many individuals will be able to comprehend that it seems to be a good message. Sending the message is not a matter of phoning up Charles X who will be on a planet near Syrius and having a chat with him about the latest ballgame results. Presumably the message is going to be some sort of encyclopedia, a distillation of much information. Given the evolution of our own intellectual processes, it would seem necessary to correlate beyond the level of just having individual articles written by individual specialists—the message will have an immense computer input, and humans will also be hooked into it. Yet the end product *will* be human.

What can we expect to communicate to beings who have evolved nongenetic ways of transmitting information about their environment? In the first chapter, Carl Sagan observed that if there is to be any chance of our contacting an extraterrestrial culture, that culture must have evolved a stable form of organization, having passed the stage of risking annihilation by playing with atomic weapons. Given such stability, we might expect the culture to last very long indeed and so might guess—and it is a completely wild guess without any foundation—that the average stable civilization is 10^6 years old. This guess is somewhat disturbing because first, it implies that we are the naive infant of the family of intergalactic civilizations, and second, it raises the question of whether we can expect to communicate with beings a million or 100 million years "more" evolved than ourselves.

Let us forget for a moment the problem of communicating with beings that have evolved in some completely alien environment and let us instead consider, given that lucky chance that man should survive for another million years, what we might have that is worth communicating to our descendants of a million years time. The changes of a few thousand years do not seem too difficult to span—given a few years to learn a bit about our society, Aristotle could probably understand it better than any of us do. But 10^6 yr into our past takes us back to subhuman hominid forms, who had not evolved speech, barely walked upright, and had very primitive tool-manipulating abilities. It is not clear that they could comprehend our ideas any better than a chimpanzee, with whom we can now hope to communicate with a few sign gestures, but with no hope of intellectual discourse about really abstract concepts.

Turning from changes over time in one evolutionary line, let us contrast the visual system of the frog with that of man to emphasize how visual systems have evolved quite differently even on planet Earth in adaptation to the environmental pressures of different ways of life. We may then ask ourselves whether a frog could communicate with a man even if by some magic it were suddenly to acquire speech.

Frogs lead rather bleak perceptual lives—they don't learn very much, and most of their lives are spent immobile, except that when they locate flies, they zap appropriately, and they will jump out of the way of enemies. In Fig. 4.5, we can compare a frog brain and a human brain. We can get some idea of scale by comparing the size of the eyes for we know how much smaller frog eyes are than human eyes, and so we can see that if the eyes were on the same scale, it would become clear what an incredibly small brain, by human standards, the frog has. The neocortex, which we hold necessary for all our higher intellectual activities, is completely rudimentary in the frog. However, what should be noted are not the gross details shown in the figure, but rather the fine cellular details in the retina, the layers of neurons within the eye. In the frog, cells respond to specific stimuli relevant to the animal's behavior; for example, frogs receive signals which might be interpreted as "there is a fly," others as "there is an enemy," etc. In the human visual system, the little work that has been done during brain surgery reveals no such "high-level coding" in the periphery. We think of frogs as much simpler than men, and yet we find that frogs have much more complicated retinas! We resolve the apparent paradox by recalling that frogs are creatures that we see today and so have evolved for as long as we have, though for a very limited environment. Thus, while our brains evolved the neocor-

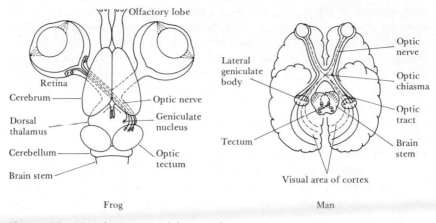

Figure 4.5 Visual systems of frog and man.

tex to allow the pooling the processing of many, many different types of information, along with such appurtenances of intelligence as memory and planning, the frog evolved better and better detectors for a limited range of stimuli important in its ecological niche. To oversimplify somewhat, we can then say that, whereas we can learn to perceive many different things in the world around us, all a frog can perceive is simple detections like whether or not a fly is present. It cannot even really perceive a fly but can only react as if a fly were there; for example, if one of two frogs starts scratching its back, then the wiggling toe is a fly-like stimulus, and the other one will start zapping at it, whereas a frog will completely ignore a dead, and thus stationary, fly. What would we have to say to a frog?

However, it might be countered that the crucial defect of the analogy with the frog is that the frog is too "special-purpose," whereas we shall communicate with "general-purpose" intelligences. With our ability to do science and to manipulate language, we can, in principle, represent *any* information. But given the evolutionary divergence we see on the planet Earth between frog and man, the evolutionary divergence in completely alien environments may be even greater. Hence, even though we try to communicate with a creature which also has a symbol system in which it can, in principle, represent everything given enough time, nevertheless, understanding from his messages what it is he is trying to say may involve an intellectual feat as great as the feat of discovering the theory of relativity or quantum mechanics. In other words, given the possible divergence between our primitives of perception, it may be as difficult to communicate our perceptions to a creature that has learned how to perceive in an alien environment, as it is to tell a blind friend what it is like to see blue. Nonetheless, the attempt to communicate about more subjective states may well have value—the very strangeness of an alien viewpoint may make us see new things, just as great poetry may succeed not so much because the poet conveys an explicit situation, but because the reader finds in the poem "chords" that change his perceptions.

What, then, can we hope to communicate with any "technological" species? It appears that purely symbolic constructs which can be reduced to a postulational system, for example, mathematics (which might include much of physics as well as even basic rules of social organization), can be communicated. Surely a preliminary phase of communicating simple axioms to educate the alien intelligence and let them educate us in basic concepts is first required. However, it seems unlikely that we could share ideas which involve affect and emotions; for example, we would be unable to communicate the feel rather than the abstract description of perceiving something.

Effective communication would require mutual education and adaptation here as elsewhere. Consider again the problem of communicating about vision with a blind person. You could convey certain mathematical descriptions involving wavelengths and also the fact that there are surfaces with different absorption coefficients, and then you could get across the idea of sensors which can use these differences to get information about distant objects which would prove helpful in navigation. But this is far different from conveying the glory of a sunset. As we have seen, even creatures which share the possession of visual systems may process their sensory inputs in radically different ways. Even though one might claim that the human brain is a universal computer—in the sense that it can achieve anything at an intellectual level given enough time—the fact remains that the difference in organization of our perceptual systems from those of other species may be such that there are certain things that are immediately comprehensible to us, but which would not be at all immediate to creatures in other civilizations.

This cautions us that even if we were to convey basic mathematical forms between the mathematicians of the two races by starting with the postulates, if the elementary components of the theory get their motivation from some sort of perception that is intricately tied to the way in which the brain which originated it has evolved in a particular type of environment, then no matter how much may be shared in the way of ability to manipulate the calculi of the system, it may still be extremely difficult to build up any sort of intuition.

Even though few of us can hope to convey fine nuances of emotion to our fellow humans in linguistic terms, we can convey much by way of empathy, because we have enough in common both in the way our brains are structured and in the type of society in which we have grown up. However, if an organism has evolved sufficiently differently, the dimensions may vary too much. You might be able to tell it some of your behavioral repertoire, but if its society is sufficiently different, it may be difficult to convey even that. Even on the planet Earth the same biological substrate of humanity has undergone much divergent cultural evolution. One finds that behavior which is considered as normal in one society—homosexuality, for example—disgusts people from other societies.

Our main hope for interstellar communication is based on the belief that a technological civilization must have numbers. It is hard to conceive of a psychology which could do technology without being able to count, and add, and multiply. To do the geometry necessary to describe the motion of its planets it must have some theory like conic sections or calculus. Thus one might expect such things to be in the repertoire of a scientist in any technological culture. Freudenthal has written a book which presents LINCOS, a language for interstellar communication. The strategy described is to start by sending messages that make clear that you are talking about numbers. You can then build towards calculus and coordinate geometry, on which basis you can start to discuss physics. One might expect (though Drake's ideas on neutron stars may suggest a counter-example) that Newton's laws hold anywhere as a reasonable first approximation, so that any scientist would eventually begin to recognize that you are talking about Newton's laws. After a while, you have a language in which you can describe the motion of particles, no matter what the senses of the creature are, or whether he perceives these motions by vision, by x-ray, touch, or another method. Just as in computers we build up higher and higher level "languages," so we may build up a higher level of language which approximates more the way we describe the world, even though we started from what we considered to be sufficiently basic physics that the alien scientist could understand. This way we can tie in more and more events to his perceptual systems, but this is still quite different from conveying the "feel." For example, the way a blind person perceives physical shapes is quite different from the way a sighted person does, and the idea of huge distances is somehow more abstract for someone who lacks a sense akin to vision.

Our emphasis on possession of technology rules out the model of a gorilla communicating with a man. The gorilla does not have language, and he does not have mathematics; it seems that we have to start at a higher level where we share at least the ability to use language. In particular, the ability to describe things that do not exist, to describe possible alternatives besides those that are actually perceived, is necessary. To build technology you have to be able to make hypotheses, to test hypotheses and, to a lesser extent, to look for unifying laws.

We need beings who can manipulate symbols in ways that are far more flexible than would result from simply going around the world taking snapshots. (Parenthetically, since we cannot assume that extraterrestrial beings will have a purely visual sense, we cannot guarantee that if you send out TV transmissions, the beings will eventually figure out how to project them on a TV screen, and then see them as we see them.) If you describe your geometry in terms of elementary number theory, then even if the being with whom you communicate can't perceive visual scenes, he certainly must have the idea of objects located in space and so can recognize a coordinate geometry description of those things. Even if he cannot really feel what you feel about certain constructs described, at least he can begin to understand your coordinatization, and he can then re-represent that in whatever is easiest for him. It might be that he will end up just putting it into a TV picture, because he does have a visual sense like ours. It might be that he will build plastic models and that people will feel them with certain senses, or some other method could be used. But the point is that you have got to have some sort of representation that is sufficiently universal that no matter what senses he has, if he can build a technology, he will have enough science to use that basic numerical representation.

Even though with our symbol systems we may be able to communicate all our mathematics, all our physics, perhaps even the basic rules of social organization, nonetheless, no matter how good interstellar communication may become, we may never know the full beauty of the x-radiation of a sunset for the inhabitants of a planet in the galaxy Andromeda.

In the preceding chapter, Arbib introduced the idea of possible intelligent and self-reproducing machines. In the next chapter, John McCarthy examines this idea more deeply and discusses the current status of studies of artificial intelligence. At present the field of developing artificial intelligence seems to be in a rather primitive state. However, modern computers, while vastly faster than the human brain, are also vastly inferior in terms of storage capacity. In addition, the picture of the outside world that can be built up by a computer, working through a television camera, is extremely crude because the pattern-recognition facilities of the human brain are vastly more developed than those of the computer manipulating areas of light and dark in a television picture. McCarthy thus believes that the field of developing artificial intelligence using computers is barely into its infancy and that vastly greater developments lie ahead. Meanwhile, teaching computers to play games gives us valuable new insights into mechanisms of human thought.

5

Possible Forms of Intelligence:
Natural and Artificial

John McCarthy *Stanford University*

The likelihood and the nature of interstellar communication would seem to depend on the intellectual nature of the beings with whom we might communicate. In the study of artificial intelligence, we attempt to study intellectual mechanisms as independently as possible of the particular ways intellectual activity is carried out by humans. Therefore the study of artificial intelligence can perhaps shed light on the problem of intelligence in the universe.

History of Artificial Intelligence

We shall begin by summarizing the history of artificial intelligence, describing its present state, and presenting some of the problems that are currently baffling specialists in artificial intelligence.

The study of how to make computers carry out activities presenting intellectual difficulties to humans really began with the advent of the stored-program digital computer in 1949.

Computer Program for Game Playing.   One of the earliest undertakings was to program computers to play games. We will examine the current levels of computer achievement in four different games. The games in question are all board games, in which players make alternate moves, but the computer programs have attained quite different levels of achievement in the four games. A computer program was written with a fair amount of effort over a couple of years to play a game called kalah. The computer now plays kalah better than any human players that we have been able to find. In fact, humans have learned to play the game a good deal better after watching the computer play. As a result of computer play, we were also able to solve the game in one of its common variants and prove that it was a win for the first player, which had not been previously supposed.

The leading characteristic of kalah which makes it possible to write a computer program that plays the game better than human players is that certain aspects of intelligence seem to be irrelevant in playing kalah. The position changes rapidly, there are no apparent strategic characteristics of positions that hold for a long time, and there doesn't seem to be very much pattern recognition involved. What seems important in the game are minor strategems. For example, if a player wants to capture a few of his opponent's stones, he will plan his strategy by trying to select from possible alternative moves. He might say to himself, "If I do this, and he does that, and I do this, and he does that, aha, I win, but if he does that at this point, I lose," and so on. That is, human beings play the game by following move trees, and the machine is much better at tracing move trees than a human. It should be mentioned, however, that the first kalah programs did not play as well as people. Two improvements in the early programs made the difference in the computer's performance. The first improvement is somewhat specific to kalah and consists of a variety of ways for determining that the game is over once one player has an unbeatable advantage. This saves much branching at the ends of the move tree. The second—called the α-β heuristic—applies to all games where players move alternately and involves a player's not examining moves that are alternatives to refuting moves. (A refuting move is one that shows that the opponent's move leading to the position is worse than one of his previously examined moves.)

The α-β heuristic is used by all human players, but it was not identified as necessary by the first writers of game-playing programs. This illustrates two facts. First, an intellectual mechanism may be compulsory for certain problems; it is difficult to imagine an effective general game player, human, alien, or machine, that did not use α-β. Second, many of the difficulties in developing artificial intelligence are the result of failure to recognize mechanisms that are obvious once they are pointed out.

The next level of performance achieved by a computer is in a more difficult game, checkers. The checker program, written by Arthur Samuel, uses a similar strategy to that used in the kalah program, and the computer also plays checkers quite well. It can beat most ordinary players, and it played with the United States Champion and got one draw in six games; in the other five the champion beat it. Since then the checker experts have learned more about how the program plays checkers, and they beat it almost every time now.

The game in which the most effort has been expended in writing programs is chess, and the best current chess program is the one written by Greenblatt at M.I.T.; this program plays class C chess, and Greenblatt thinks he can get it to play class B chess by an extension of his present methods. Several rival programs have recently been developed, and a computer chess tournament has been proposed. Like all good chess players, Greenblatt is coy about whether he will participate. He is interested in playing his program mainly against human players. In fact, computer programs have been admitted to the U.S. Open Tournament provided they obey all the rules including the time rules.

Chess programs involve much more than simple tree search, which yields a very bad program even with α-β. Much of the program is specific to chess and includes techniques for recognizing threat situations, techniques for evaluating positions, and techniques for determining the effects of complicated exchanges of pieces. Of more general significance is the computer's ability to compare the potential of a move with the requirements of a position; for example, if in a certain variant being considered, one is down a bishop, one should not engage in plots to win a pawn.

Jonathan Ryder, a graduate student at Stanford, is writing a program to play Go. The program still plays extremely badly, and this is because the intellectual mechanisms that we understand and know how to make a computer carry out and which work fairly well in chess, are very weak in Go. The problem in Go is the large number of moves. Since a move consists of putting a stone on a 19×19 board, there are 361 possible first moves, 360 replies, and 359 replies to that move and so on. Therefore, the method used in chess, that of following out a move tree, immediately fails in Go because the tree will be $361 \times 360 \times 359 \ldots$ moves in size, and Go players sometimes look a fair distance ahead. We examine this strategy and ask what is wrong; a human player certainly doesn't look at all those alternatives. Then we discover that the human player divides the board up into regions on the basis of the stones that are already on the board and considers the regions separately. To figure out whether a group of stones is safe or is in danger of being captured by the opponent, the human player does a local analysis. He goes through a move tree, but the move tree is limited to moves in the immediate area and includes only those moves which are suggested by certain principles. He may come to a conclusion that the group is safe only if he moves, but if he doesn't move, it will be captured, and he will lose a certain number of points. He will remember this while he thinks about what his opportunities are on the rest of the board, and then after a while he will begin to think about interactions between these local situations. He will note that a move at a certain place affects both situations and so forth. We do not fully understand yet how to program the recognition of local situations. In chess, local situations also exist, and maybe we can't get much beyond the level that Greenblatt's program currently achieves without taking this into account, but in Go, their recognition is essential for even moderately good play. Ryder has had some success in this direction.

Computer Theorem Proving.   Another area of intellectual activity recently undertaken by computers is that of proving mathematical theorems. At first this work was done in a large number of different formalisms, but now most of it is done in Robinson's resolution formalism of the predicate calculus. There are quite a number of theorem-proving programs that work by resolution, and quite a number of different kinds of problems have been formalized in predicate calculus.

The most spectacular result of computer theorem-proving was the solution of a known unsolved problem in lattice theory. A mathematician looking at computer output noted that one of the computer-proved formulas implied the resolution of the conjecture directly. Moreover, lattice theory seems to be particularly suited to computer theorem-provers because much of it involves formula manipulation; complicated structures of lemmas and the use of examples are less essential in lattice theory than in other branches of mathematics.

Computer Interaction with the Real World.   Much work has also been done in the area of computer interaction with the real world. For example, a computer is equipped with an artificial arm and an artificial eye (a television camera). Programs can then be written to assemble objects out of parts. It turns out that the most difficult problems are in the area of vision since they involve going from a TV image in the computer (in our case a 256×333 array of four-bit numbers) to a list of the objects in the scene with descriptions of their positions and attitudes so that the programs that control the artificial arm will be able to know where to reach out, grasp, pick something up, and move it.

At present there are several approaches to the problem, all of which are clearly quite limited even in their potential accomplishment, let alone in their present accomplishment. One approach involves the use of scenes that are composed of objects with flat faces. Thus there are definite edges between the faces, and there exist programs that can find these edges and even distinguish edges of objects from edges of shadows, which is not a trivial accomplishment, since sometimes the shadows are more prominent than the edges of the objects. In any case, block-stacking is possible; a computer can pick up a batch of blocks and make towers out of them, and we hope to be able to do some more ambitious things shortly.

Another approach being followed at Stanford Research Institute involves recognizing objects in a different way. The program divides the region to be examined into a large number of subregions, for example, 100×100; it first characterizes each subregion, and then it joins together adjacent subregions that have the same characterization, for example, subregions of about the same color and the same shade would be joined if color or shade are significant. Thus it builds larger regions and finds the boundaries of the regions, then it groups these regions together to make objects. This method has some potentiality for dealing with curved objects. Still other possibilities are being pursued.

Current Research Problems

We can complement the discussion of the concrete accomplishments of artificial-intelligence (AI) research by examining the current research problems. In the author's opinion, there are two main classes of problem in AI research. The first is the discovery or invention of programs. Most of the mechanisms are discovered in human behavior—more often by introspection than by formal psychological experiment. However, sometimes mechanisms are invented that have no obvious counterpart in human or animal behavior. An example of a discovered mechanism is $\alpha-\beta$, and associative memories based on hash-addressing is an example of an invented one. Mechanisms are often found by examining the reasons for the disappointing performance of a program containing mechanisms previously thought adequate to solve a certain class of problems. Frequently, the new mechanisms, once found, are considered obvious. Particular intellectual mechanisms are often called heuristics, and their study is called heuristic programming.

The second class of problem is more basic, but this fact has only recently become apparent to many workers in the field. These are the problems that involve questions of what an intelligent being, human or machine, can know about the world and how this information should be represented in the memory of the computer. In the early game-playing and theorem-proving programs, it was possible for the programmer to devise an ad hoc representation of what was believed to be all relevant information. Present programs for game-playing are all based on these ad hoc representations of positions. Any strategic or tactical concepts are represented by features of the program. When we want to design a program that approaches the human generality in reacting to real-world situations that include observing the physical world, receiving information about it expressed in a natural language, deciding whether sufficient information is available to take successful action in a given situation and, if not, deciding how to get more, then the ad hoc representations are inadequate. Then we must equip our program with some ideas about what the world is like in general (metaphysics) and some ideas about what knowledge is available and how more can be obtained (epistemology). The problems are those of the above-mentioned branches of philosophy, but when we look at what the philosophers have done, we are disappointed. Almost everything that they have proposed is too vague; for example, we cannot program a computer to look at the world in the way recommended by Wittgenstein. Also, much philosophy seems clearly wrong; the recommended ways of getting information (to the extent that they are precise) just won't work. The positivist philosophers seem to have thrown out real problems in their efforts to clean out meaningless ideas.

Recently, a beginning has been made in developing a formal language capable of expressing what human beings know and robots need to know about real-world situations in order to take successful action. This language has involved expressing, in first-order logic, descriptions of situations and the effects of taking actions in them. The desire has been to formalize the situation sufficiently well so the fact that a certain strategy is appropriate to realizing a certain goal is a logical consequence of the description of the particular situation and of the general information about the effects of different kinds of actions. Moderate success has been achieved, and the research continues. We hope the results will be of philosophical interest as well as of interest to artificial-intelligence researchers. Within the last two years, a new formalism, based on Hewitt's language PLANNER, has been developed by Hewitt, Winograd, and others at the M.I.T. Artificial Intelligence Laboratory. This formalism represents much information as procedures and has been particularly effective in translating information originally expressed in natural language into computer form. It also gives a new approach to expressing generalizations that have exceptions, not all of which can be presented along with the original generalization.

The epistemological problem is also acute in the development of programs that physically manipulate the world on the basis of visual information.

Possibilities for Interstellar Communication

Research on the development of artificial intelligence is relevant to the problem of interstellar communication. The first conclusion to be drawn from the preceding discussion is that the mechanisms of intelligence are objective and are not dependent on whether a human being or a machine or an extraterrestrial being is doing the thinking. To play chess well, certain processes, for example, processes of search and processes of factoring a situation into its subparts must be carried out, and it would seem that these processes are independent of the intelligence carrying them out. If a person wants to discover physics for himself on the basis of experiment, again the intelligence procedures that must be followed seem to be determined to a large extent by the nature of the problem, rather than by the nature of the learner. This suggests that we should expect to find other intelligences in the universe using procedures similar to the ones which we use and which we would like to program our machines to use. Of course, they may think slower or faster than we do, and they may have pursued certain areas of knowledge to a lesser or greater extent than we have.

There are much greater possibilities for difference in motivational structure. Most computer programs are not properly said to have motivational structures; they just run. However, many of our attempts to make intelligent programs involve interpreting certain expressions in the computer memory as goals. The program compares the present situation with the goal; this comparison suggests possible actions, the predicted results of which are also compared with the goal. Subgoals are generated subordinate to the main goal, and these are pursued. It seems that such programs may properly be said to have motivational structures, and we may also try to interpret human behavior in terms of formal goal structures. However, the programs we have written so far and even programs that we contemplate writing in the future to achieve our purposes seem to have simpler motivational structures than human beings have.

Human beings often change their goals in a stronger sense than programs change subgoals in pursuit of a main goal. In fact, a human does not have a main goal in the sense of some function that he conducts his life in order to optimize. We often find ourselves in states of relative goallessness, where what we want is to find a goal.

To consider another example, a dog's motivational structure can be described approximately in the following way: sometimes the dog is hungry and wants to eat, sometimes he is thirsty and wants to drink, sometimes he is driven by a need for sex. When none of these needs are unfulfilled, he is content to lie down and rest. One could imagine that some very intelligent organism could have a motivational structure similar to that of the dog. It would solve problems as they present themselves, but it would not do much when no problem presents itself. On the other hand, other aliens might be motivated by curiosity, they might want to find out as much about the universe as a whole as possible, which they would do by pursuing science and also by exploring the universe. We can imagine beings motivated by a drive to expand; they might want to convert as much of the universe as possible into their own substance. It seems that all of these are possible stable forms of motivational structure for an intelligent being to possess, though it is not so easy to imagine which of them could have evolved.

With regard to social organization, there are also many possibilities. Thus we tend to presume civilizations composed of many independent beings with distinct individual goals interacting with each other, but this is not inevitable. Science fiction writers have imagined a single intelligence that has incorporated more and more material of its original planet into its structure and which has goals of expansion or curiosity.

Some ideas about the communicativeness of extraterrestrial intelligence can be obtained by speculating about our own future. Of course many people are trying to plan our future for the next 100 or even 1,000 yr, but it seems that these plans have a certain arrogance. Suppose that we look back and ask what attention we pay today to plans made for us, especially in the technological and economic area, by people 100 years ago. It turns out that we are not even very curious about what they thought the future would be like, and we can assume that our descendants will be similarly uninterested in what we predict and plan for them. Thus it seems that planning should be limited to one doubling time of our technology in the area in question. However, speculations about our own future may be relevant to other technologies that we might encounter. Some of those questions concern artificial intelligence itself. A question frequently asked is whether we will develop computer programs that are more intelligent than we are?

At present it appears likely that we will develop computer programs that are more intelligent than we are unless we decide not to for some reason. We cannot predict a date, because there are some important problems that have not been solved. The situation may even be as primitive as the space program before Newton. Nevertheless, there seem to be no limitations on machine intelligence short of human intelligence, and if we can construct a computer with the human level of intelligence, we can get much more intellectual work from computers than from our whole race just by using faster computers and larger memories. The question of how we will use this artificial intelligence remains to be answered. Will we limit it, will we in some sense merge with it as has been proposed in some science fiction stories in which a human being adds to his own intellectual and physical powers? Just as our cells are replaced every 7 yr, after some time of interaction with these additional powers, the intelligence would have to be described as resident mainly in the artifact and only slightly in the original carcass. It is very difficult to predict what we will do. However, imagine that we either directly or through machines acquired some much higher level of intelligence. Then we would have to ask the following questions: Is the universe still interesting to us, that is, does it still have structure yet to be discovered, or will we discover the fundamental structure of the universe once and for all? Another question concerns exploration. Is it true that when you've seen one galaxy you've seen them all? Are all high intelligences alike, or are they sufficiently different to be interesting to each other? In a sense, two computers are not interesting to each other—they have very little to say to each other unless thay have substantially different data bases. Any computer can run the programs that any other can run, though they might have different data sets associated with them, but there does not exist the same reason for mutual cooperation among computers as there does among human beings.

To summarize, in this paper the point has been made that artificial intelligence is possible and that intelligent entities in the universe may be similar in their intellectual properties because the methods of intelligence are determined by the problems. However, intelligent entities may differ considerably in their motivational structures. The possibilities for interstellar communication depend partly on this factor and partly on presently undecidable objective questions that determine whether intelligent entities will have much of mutual interest to say to each other.

In his formula in Ch. 1, Sagan presented us with the factor f_c, which represents the fraction of planets with intelligence that develops a technological phase during which there is a capability for an interest in interstellar communication. This factor thus touches not only upon technological capabilities, but also upon the attitudes of a society.

In the following chapter, S. Aronoff goes beyond the questions of chemical and biological evolution discussed in previous chapters to discuss social evolution, which involves consideration of the goals of social organizations and the commitments of resources which such organizations may be willing to make to meet those goals. The recent history of mankind is marked by the rise and fall of separate civilizations around the globe. However, our civilization has presently become truly global, with all parts of the world interacting simultaneously with all other parts. True, in some places individual nations are very powerful, and in other places they are very weak, and these power balances may shift from one place to another within the overall global community. Are we thus beyond the rise and fall of global civilizations? Can the complex global interactions, once established, be eliminated without complete global destruction? It is upon this question that the longevity of our civilization depends. This longetivity L is the last factor in Sagan's equation.

6

From Chemical to Biological to Social Evolution

Samuel Aronoff   *Simon Fraser University*

Introduction

Scholars in the theory of evolution have suggested, from time to time, that terrestrial biological evolution is drawing to a close and that a new phase, called *social evolution*, has begun. In its most gross terms, this would mean that primordial matter has progressed from its initial form as hydrogen, through the synthesis of the various heavier elements in stable star formation, through the complexities of *chemical evolution* on the surface of the earth, followed by *biological evolution*, and that now, after approximately three-and-one-half billion years of biological evolution, it is rapidly entering a third era, that of *social evolution*.[1]

This concept immediately raises a number of questions: (1) Is the line between biological and social evolution drawn so sharply that whereas previous geologic major changes are discussed in terms of at least thousands, if not millions, of years, we are now able to think in terms of tens of years? (2) Rather, is the question not a general one, but concerned essentially with human beings, whose total generic span is not much greater than a couple of million years? (3) Is the question narrower still, namely that we are witnessing a grand change in human life-style, so that *civilization* as we have known it is being changed, but that the world as a whole would be no more affected than by the passing of, or gross changes in, any of the important genera of fauna which now inhabit it?

In the following discussion we will speculate on the general theme that at least that aspect of the theory raised in the third question is pertinent, quite possibly that raised in the second, and, with much greater reservations, even that raised by the first.

On the Conditions for Transition from Chemical to Biological Evolution

The framework for a general theory of biochemical evolution is now being established, especially with regard to the statistical mechanics of the transition from chemical to biological evolution. This has meant, in particular, the clear definitions of "survival of the fittest" in molecular terms (Allen, 1970), of the principles of self-assembly of macromolecules (Eigen, 1970, 1972), of the theory of hierarchies of matter and their relation to irreversible thermodynamics (Prigogine, 1970, 1972), and of the statistics of evolution of nucleic acids (Kuhn, 1972).

As a consequence of these and adjunctive aspects, the logic of the transition from chemical to biological evolution as an inevitable act is now hypothesized. In other words, the formation of living things may be a natural consequence of the evolution of matter.

We assume, in consonance with much experimental endeavor, that the prebiotic oceans contained essentially all of the monomers common to modern organisms. These are presumed to have arisen as a result of the interactions of various sources of energy (ultraviolet radiation, heat, lightning, wave action (cavitation), meteorite impact, etc.) with the primeval

[1] Cultural evolution is sometimes described as an intermediary stage between biological and social evolution.

(secondary) atmosphere of the primitive earth.[2] These monomers were not made from the primeval atmosphere (p.a.) in the main, but from other monomers (which, of course, were themselves either derived from the p.a. or from monomers resulting from them). In other words, compounds arose for the most part by virtue of the development of *chemical systems,* in some of which certain of the monomers were able to act as catalysts. (For example, cyanide ion which is known to act catalytically in decarboxylation and various lyase syntheses, parallels the action of thiamine, where the action of the thiazole ring may be depicted as cyanogenate in catalysis.) It is then possible to define as *living, any chemical system with the capacity to evolve indefinitely by natural selection* (J. F. Cross). The notion of a living system is distinguished from that of an *organism* which is described by G. Allen[3] as *"a (multimolecular) form of life whose reproduction depends upon a copying mechanism."*[3] Allen then further defines "indefinite evolution by natural selection" as involving (1) self-dependent multiplication, and (2) mutation. The first is any process in which molecules are produced at a rate positively dependent on their number, and the second is any system furnishing potentially unlimited variation, as a result of which new structural features may be acquired. (Allen provides models illustrating the principles enunciated.)

It has been shown by Prigogine[4] that open systems may develop states in which the concentration of one or more of the constituents may have a periodic value. Further, under conditions where this may be visualized (for example, by use of a dye) standing waves may be observed, which then represent positions of equal concentration of the visualized constituent(s). These standing waves are then equivalent to a new ordered structure of the system or, as Prigogine prefers denoting it, a *new state of matter.* Indeed, under some conditions, these standing waves may serve as foci for the construction of branch systems, that is, still higher echelons of structural hierarchy. Eigen[5] has examined the parameters involved in

[2] A gross list of comprehensive texts on the "Origin of Life" would include the following:
Buvet, R. and C. Ponnamperuma (eds.), *Molecular Evolution I Chemical Evolution and the Origin of Life,* North-Holland Publishing Company, Amsterdam, 1971.
Calvin, M., *Chemical Evolution Molecular Evolution Towards the Origin of Living Systems on the Earth and Elsewhere,* Oxford University Press, New York, 1969.
Florkin, M. (ed.), *Aspects of the Origin of Life,* Vol. 6 of *International Series of Monographs on Pure and Applied Biology: Modern Trends in Physiological Sciences,* P. Alexander and A. M. Baog, Pergamon Press, Oxford, 1960.
Kenyon, D. H. and C. Steinnan, *Biochemical Predestination,* McGraw-Hill Book Company, New York, 1969.
Kimball, A. P. and J. Oro, (eds.), *Prebiotic and Biochemical Evolution,* North-Holland Publishing Company, Amsterdam, 1971.
Oparin, A., *Genesis and Evolutionary Development of Life,* Academic Press, New York, 1968.
[3] G. Allen, "Natural Selection and the Origin of Life," *Perspectives in Biol. & Med.,* Autumn 1970, 109–126.
[4] Prigogine, I. and G. Nicolis, *Quart. Rev. Biophys.,* **4** (1971) 107–148.
[5] Eigen, M., *Quart. Rev. Biophys.,* **4** (1971) 151–209; *Naturwissenschaften,* **58** (1971) 465–523.

self-organization of molecular systems, especially with regard to regulation to produce stability, mutability, and selectivity. Recently, H. Kuhn[6] has described the physical requirements for the modeling of the nucleic acids and their functions to become part of a living system, especially to show the statistical plausibility of ordinary physical chemistry in providing polymeric conformations capable of natural selection.

Biologic Evolution

The concept of entropy is frequently involved in discussions of living systems, and it has frequently been stated that such systems appeared to act in violation of the second law of thermodynamics. This argument generally took the form that living systems developed higher-order systems (themselves) out of relatively chaotic ones (their food). In other words, organisms reversed the increase in entropy expected of spontaneous chemical reactions. Prigogine and his colleagues have elucidated with classic simplicity that the variation in entropy (dS) during a period of time (dt) consists of two parts, that part due to exchanges of the system with its environment (d_eS) and that part due to irreversible reactions within the system (d_iS), such as chemical reactions, heat production, etc.[7] Although the latter term is always positive (that is, $d_iS > 0$), the first may be either positive or negative. Consequently, if the organism is able to transfer entropy from itself to its environment, it may reach a state in which the steady-state entropy *within the system* is less than the formal entropy entering it. (Thus Prigogine develops the cognomen of *dissipative structures* for ordered systems arising under such conditions.)

While the concept of entropy is generally associated with the number of degrees of freedom of a system (that is, its ordered structure) it may also be interpreted in terms of information theory.[8, 9]

Assuming, then, that one may correlate biologic evolutionary levels with information content of organisms, one may then ask what relation, if any, this has to do with survival value. Clearly, there are many examples of organisms which have persisted for literally eons of geologic time with relatively little apparent change and with obviously less information content than many families which have evolved since then. The question is what constitutes ability for an organism to survive with less information content compared with one having higher information content. The answer to the question would appear to be that each species of organism samples its own environmental niche from the total environment; that which constitutes an ecologic niche for one may have no bearing on

[6] Kuhn, H., *Angew. Chem.* (international ed.) **11** (1972) 798–820.
[7] Prigogine, I., G. Nicolis, and A. Babloyantz, *Physics Today* (Nov. 1972) 23–28.
[8] Theodoridis, G. C., and L. Stark, *Nature,* **224** (1969) 860–863.
[9] Brastow, W., *Science,* **178** (1972) 123–126.

or relevance to that of another. The necessity to compete for survival arises only when and to the extent that niches overlap. Thus the strongest competition will arise between very closely related organisms, for example, those related by single-point mutations. What is information for one is noise for the other.

However, the interaction between a species and its environment is not a one-way street. When a diamond wheel cuts through wood, a small fraction of the diamond edge is abraded. Thus adaptation, by mutation and selection, may involve not only a fitting of the organism to the environment, but to varying extents, the converse is also true. In this way an organism may more perfectly fit an ecologic niche, in part because the niche has been adapted to the totality of characteristics unique to that species of organism. Every species alters and, in turn, is altered by the environment. In this way all species become part of the historical process. When the information content of the reconstructed environment is completely utilized by the organism, so that no additional information is obtainable with its genetic structure, it is then perfectly adapted, that is, it is at equilibrium with its environment. (By definition, then, it is able to survive under the conditions of its environment, since its predators constitute part of the environment.) No subsequent mutation would be of value to the organism; the potential for evolution would cease, and all things remaining equal, evolution would cease.

But there are equilibria and . . . equilibria, that is to say, an equilibrium may be reached with respect to one parameter prior to its having been attained with respect to another. (As an example, in equilibria involving isotopes, mass equilibrium is generally attained in times considerably shorter than isotopic equilibrium.) Thus in terms of evolutionary pressure, equilibrium may be reached, but this is not necessarily (and usually will not be) identical with optimal processing of information content within the species. The partitioning of the information content and response process within a species in reaction to its environment is called *social evolution*. In theory, then, this may occur in any species at or near an evolutionary equilibrium.

The Nature of the Process of Social Evolution

The struggle for survival is an integrated process. In the following discussion we shall compare one "wild-type" species with another. However, unless the population is clonal, there is a distribution, generally Gaussian, about each determinant. In the general struggle for survival the distribution of information content among the several parameters is less than optimal. The process of social evolution then consists of the redistribution of the information process among the several parameters in such a way that one or more of them is optimized. In most lower species of animal this appears to take the form of a chemical vector, resulting in morphologically distinct classes of individuals, as with the ants or bees, for example. The objective, however, is not structural, but rather functional, and thus so long as there is some ready device for recognition of the functional assignment within the species, the pressure for structural identification is not imperative. Thus, in higher animals social structure may be revealed by the geometric position of the leaders within the area of operation; in humans there are many added subtleties, such as manner of speech, clothing, and so forth.

In a system at equilibrium, this partitioning will result in correspondingly stable subclasses within the species, but in systems at quasi- or near-equilibria, the introduction of a new internal or external parameter into the process may result in strong and/or sudden redistribution of entropy within the system. In general it is not feasible to make detailed generalizations. As in a chemical reaction, where it is not possible to delineate the reaction mechanism without precise awareness of the chemical nature of the reactants and products, so it is not possible to optimize the lattice of a social model without intimate knowledge of each of the lattice points, that is, the kinetic factors interrelating the social parameters.

Homo Sapiens and Social Evolution

It is unreasonable to compare the social evolution of insects with that of humans for two reasons: first, the comparative lengths of time for the evolutionary process to occur in each is different (about 200 million years in the case of the insects and possibly somewhat over 2 million years in the case of *homo sapiens*). Second, there are differences in the relative complexity of the two forms of life, as measured by the estimated numbers of brain cells in each (approximately 10,000 in the case of ants and about 10 billion in humans). Indeed, when one considers the number of parts in the average auto (on the order of 3,000 to 5,000) or, more properly, the engine and its associated parts (about one-tenth this number) with that of the ant, it is remarkable that an animal with as few brain cells as an ant (that is, with only 20 to 40 times the information content of an internal combustion engine) has been as successful as it has obviously been. (This comparison assumes that each brain cell constitutes only a single bit of information and neglects intercellular connections and the resulting integrative aspects.)

We now know that most aspects of human culture, previously believed to be unique, are not, whether they are the use of tools or complex forms of communication. We do not even know, for example, whether the mastery over fire was available to species of man earlier than Pekin or Swanscombe (about 125,000 years ago) or whether the Cro-Magnons of 40,000 years ago, despite their obvious skill at art, had yet developed the rudiments of a written language. The development of agriculture, apparently coincident with the formation of towns and cities, appears to be little more than 10,000 years old, and there is no evidence to suggest that it is more than 20,000 years old. The the most of the cultural factors commonly associated with the emergence of *homo sapiens* are extraordinarily recent, covering at best only 1/1,000 of the known existence of the genus, though admittedly a much larger fraction of the existence of the species.

Social evolution in historic man is therefore not an unreal objective for consideration, despite the brevity of its time span. This is manifestly reasonable if history is a record of human activity, since it must then follow that the amount of history developed is simply a function of the number of people, and there are now alive virtually as many humans as have lived in all of human prehistory.

This recorded period of history has witnessed the rise and fall of a number of civilizations, analyzed with some detail and attention to philosophical implication by Spengler[10] in his well-known work. Whether one agrees with his general conclusions concerning the periodicity of civilizations and his application of these conclusions to the present civilization, it is pertinent to ask whether we are indeed now witnessing the demise of a civilization and the transition to a new one, or whether the changes now occurring are merely fluctuations in man's general progress through time. (We recognize that the generality of the word "civilization" is being abused, that there are a variety of *cultures* extant, ranging from Stone Age, through Early Agriculture, to that of modern society which is now so homogeneous that it is very difficult to distinguish one country or one major city from another.)

The major argument of this presentation is that we are indeed at the end of a cultural period and on the threshold of another. The argument may be divided into three categories: (1) the acquisition of unlimited sources of energy; (2) the acquisition of detailed self-knowledge; and (3) the unification of human needs.

The Acquisition of Unlimited Sources of Energy.   The present usage of power by each person in the United States is about 300 BTU/yr (equivalent to approximately 12½ tons of coal or 10 kw thermal). The terrestrial average is thought to be about one-seventh of this quantity. Weinberg and Hammond[11] calculate that at present usage rates, the fossil fuels will last no more than 30 yr in a population of about 2×10^{10} (20 billion), assuming that the individual requirements will double to 20 kw/person/yr. Thus it has become necessary to consider alternative means for developing energy sources. Of all the theoretically feasible sources (hydro-, tidal, geothermal, and so forth) the only ones which would appear to

[10] Spengler, G., "The Decline of the West," Alfred A. Knopf, New York, 1962.
[11] Weinberg, A., and R. P. Hammond, *Am. Sci.*, **58** (1970) 412.

have practical feasibility within the foreseeable (that is, technologically immediate) future are those of solar or nuclear energy. (Some of the other sources, such as hydroelectric energy, are quite feasible but by themselves are capable of assuming only a very small fraction of the total needed.) Solar energy is at present extremely inefficient, and the feasible choices appear to lie (except by the lowering of living standards) in the nuclear-energy field. Here the major problems are those of heat dissipation and disposal of radioactive wastes.

While the immediate technology permits only fission reactions, there is reasonable confidence that breeder reactors, either of the deuterium type or the catalytic burning of thorium (^{232}Th) or uranium (^{233}U), can be developed. The large amount of deuterium in water (1 : 5,000) makes the supply virtually inexhaustible. In the case of thorium, the Conway granites, containing 30 ppm Th, mined at twice the present rate of coal, could provide the required 500 g/day of Th for only 200 yr. However, there is enough lower grade granite (containing 12 ppm Th) to last literally millions of years. Thus, if these technologies are resolved, a supply of energy is available which is virtually inexhaustible for a level of population foreseeable within the next half-century.

The Acquisition of Detailed Self-knowledge. The development of the physical sciences in the last quarter century has affected progress in biology in three diverse ways: theoretically, structurally, and analytically.

Biological sciences have been affected by the profound and detailed application of physical chemistry to biological compounds, especially the previously intractible polymers and their homogeneous and heterogeneous aggregates. For example, whereas the properties of proteins were mysterious and wonderful only a quarter century ago, the growing understanding of the interaction of the various forces (ionic charge distribution, hydrogen bonding, hydrophobic bonding, dispersion forces, and so on) resulting in growing understanding of the unique conformations in specific solvents, has dispelled much of the mystery. The same can be said of the other common polymers: the nucleic acids and polysaccharides, and the heterogeneous conglomerates. That man can at will dissociate and reassemble (and, if necessary, biosynthesize) the major protein-synthesizing organelle, the ribosome, is truly a profound inroad into the detailed knowledge of life.

The second aspect of the effect of the physical sciences on biology is the elaboration of highly sensitive and unique instruments and techniques for the physical investigation of organic compounds, such as the sophistication of x-ray and neutron analysis for the absolute determination of structure, the development of magnetic-resonance spectrometers, the mass spectrometer, single and dual spectrophotometers, the spectropolarimeter, the varied forms of chromatography, and so forth. The development of these instruments and techniques has resulted in a renaissance in biology, changing biochemistry from essentially a study of small monomeric bio-organic compounds or of nutrition, to essentially an investigation of the physical chemistry of biological systems, the chemistry of cellular and organic interactions, the biochemistry of development during growth, and probably most importantly, it has resulted in a strong insight into the chemistry of heredity and thus of biologic change.

The third point relates to the entrance into biology of the physical scientists themselves, the purveyors of both the theory and the instrumentation, imparting an academic rigor to the field, putting substance into the wispy dreams and the descriptive tongue of the old-style biologist. Biology is becoming analytic—not so much in the mathematical sense as in the chemical; it can be analyzed, and its activities can be depicted in physico-chemical models. We do not lose sight of the philosophical as well as practical arguments of the systematists vis-à-vis the reductionists. However, if we can dis- and reassemble ribosomes and mitochondrial electron-transport systems and various types of membrane assemblages, if we can indeed construct the subcellular components of cells, there is surely no barrier to the construction of cells. We are now aware of totipotency in both plant and animal cells, that is, given a single cell we can reproduce the entire multicellular, differentiated organism. The possibility of purposeful construction of higher organisms, including man, must not be overlooked.

This last development is not in the immediate present, but (like the vista of virtually unlimited sources of cheap energy) it need not be the distant future before it becomes part of the human enterprise. In the ultimate, we shall have an intimate knowledge of the human cell, how it develops into a human being, and how a human operates. When we have reached that stage, we will have learned how to manipulate man's being, his physical structure, and his mind. Our exodus from the Garden of Eden will have become complete.

Development of the Social-Evolutionary Model

There is now general acceptance that within a comparatively few generations—three, five, or ten—human beings will have saturated the earth. The saturation will occur not so much in terms of physical contiguity as in the restriction of one or more of the essential human resources (air, food, water, shelter, energy, metals, or fibers).[12] Platt and others [13] have given reasoned but forceful arguments on the various problems and resource limitations confronting humanity and their estimated time limits before becoming catastrophic. There is current debate concerning particu-

[12] Weinberg, A., and R. P. Hammond, *Am. Sci.,* **58** (1970) 412.
[13] Platt, J., *Science,* **166** (1969) 1115.

lars and the validity of some of the assumptions, but there is no doubt that limits exist to one or more of the major human needs, that these limits will be reached "soon" from an historical point of view, and that they must inevitably impose strictures on human activities. As a consequence, one may assert that the fate of an individual is now so inextricably interwoven with that of his fellow men that the future of one is the future of all. A common human mode is being developed and social heterogeneity is giving rise to homogeneity.

We are witnesses not only to the near-depletion of critical resources and mutual congestion of various kinds, but to a contraction of time and space, so that communication of any kind, whether verbal, epidemiologic, economic, or political, is virtually instantaneous and ubiquitous. History is being written with a rapidity and certainty that it has never previously witnessed. For better or for worse, the unity of the world is becoming a fact. The old civilization is dying, and a new one is being born.

In summary, as the overwhelmingly dominant biological species on the earth, we are (or shortly will be) in a position to control the destiny of all others. We have developed a civilization which is highly universal and, because of technologic innovations, critically interdependent. Further, we are slowly creating a completely synthetic environment. Most important of all, we appear to be theoretically capable (at some not-too-distant future time) of directing the potential of the species. As a consequence, one must conclude that our period of biological evolution is drawing to a close. Life as we have known it will be changed profoundly. We are entering a new period of social evolution. The period of social evolution initiated by the domestication of man in the pre-historical city-state is ending rapidly.

Consequences

What about it? First, we need to understand that social evolution is not a doomsday phophecy: there is no reason for gloom. On the contrary, man will be able to control his destiny in a manner which the random process in biological evolution did not permit. If man is wise and learns his nature, life should be most enjoyable for all.

Second, one should not be frightened by its synthetic aspects. We consume prepared breakfast cereals or other foods made using primitive homologies and are not frightened, nor do we recoil at the use of synthetic fibers for clothing or of plastics in much of our technology. Man-made forests can be much more pleasant than those developed *au naturel*. Our fears are born of ignorance; our pleasures, of teaching. We enjoy that which we have been taught to enjoy and which is harmonious with our nature. There is no need to fear the future if we are wise. We can redo the earth to our pleasure.

The assertion that a civilization is nearing its denouement should occasion no surprise. Other earlier civilizations have arisen by virtue of some technologic advances (for example, the reduction of iron or the development of new concepts in agriculture, such as irrigation systems). Each civilization in general brought something novel and was, in itself, an aspect of social evolution. But the fraction of the human race on which it was operating was appreciably less than the majority, and thus it was effective only by social diffusion. The period when mankind ceased being a hunter and nomad wandering through the environment like the other beasts of nature and began establishing his own environment marked the beginning of his social evolution. This evolution continued through several specific civilizations as an incremental diffusive process, primarily as the development of the city-state, and culminated in highly interdependent technology, limitations of natural resources, and tremendous contractions of time and space.

The new phase that has begun will thus be a major phase of social evolution. One can envision a series of developments, most of them awe-inspiring, in the transformation of the environment. (The argument that man *need* not change the earth, that he can maintain a pseudo-steady state in the present system is contradicted by the empiricism that nature is never static.) Having command over the earth, man *is* gradually transforming it. Assuming his survival by mutual agreement not to permit genocide, the only question is one of rate of transformation. The more tightly the human race is associated, the less it will suffer the insults of change, as there will be a natural tendency to leave less and less to chance.

Among the obvious endeavors will be a gradual transformation of the earth's surface according to man's desires, consistent with the knowledge of the forces operating above and within the earth. Energy must become common and relatively cheap, as water is today. Food will be for the most part "synthetic," the raw materials being grown under highly controlled conditions. Human services must be as automatic and guaranteed as we expect our public utilities to be today. In other words, we expect that human needs will be filled routinely and automatically.

Technology may be so developed as to exploit the solar system for specific, minor elements that are rare or dissipated on the earth, for storage, or for experiments. Eventually if the velocity of interstellar travel becomes sufficiently rapid—an extremely remote possibility—man can extend his voyages beyond the solar system to at least the nearby stars, as the relativity factor will then permit these voyages to occur in a reasonably short span of time.

Utopian as the above predictions appear, they are only the first steps in the new phase of human social evolution. It is not logical to expect planning and administration of the earth without common consent and participation. Whether man will survive the challenge to common purpose and control is not the question; he simply cannot survive without it.

Once this becomes a fact, once the grandiose strategies of technology have begun, then the more profound problems will surface. What will be the nature of the social order? How will man be reproduced? Which genes should be modified, by whom, and how will it be determined if such modifications are useful socially? Will there be attempts at genetic segregation? The list goes on and on. There are no models which man can use; empiricism is the only basis for decision.

The real problems, it would appear, are not in the technologies involved. These are games—serious games—which men play with nature. But the question is whether it is these very games which will constitute the driving force in social evolution (that is, the necessity to avoid the static, closed system of the earth and its immediate environment) or whether, if we are earth-bound with occasional forays onto dead but nearby planets, we devote our energies, as man has done before in times of leisure, to the arts. On the other hand, it is difficult to believe that a closed system of this type can avoid social stratification, especially by genetic programming. One may argue that this in itself is not necessarily immoral. Those who are programmed for a particular kind of existence would probably be just as content with their lot as others for a different one. However, should this become the case, then we must admit to the end of the game. Life itself might still go on, but social evolution would have ceased.

The alternative point of view is not improbable and is far more intriguing; namely, that man will be able to develop his space technology for interstellar travel and that his horizon will be the universe—a gradual diffusion with adaptation to new ecologic niches in other stars, to interactions with other life forms through the inevitable contacts on other planets, to an even greater understanding of nature in all her majesty.

How to Communicate

We now pass from discussions relating to the factors that are needed to try to estimate the number of intelligent civilizations that may exist in our galaxy, to chapters that deal with the problems of communicating with those hypothetical extraterrestrial civilizations. The next chapter, by R. N. Bracewell, treats the problem of "finding the needle in the haystack." If you are trying to establish communications with someone, and you think that he may be trying to establish communications with you, but you do not know about which of any of several billion stars his planet may revolve, then you certainly have a needle-in-a-haystack problem. How do you know that he is sending messages in your direction when you are pointing your telescope at his star, or vice-versa? Can you even be sure that his physics is not much more advanced than your own, and that he has not discovered some channel of communication involving principles of physics of which you are completely unaware, and which he is expecting you to use in attempting to contact him?

Bracewell's solution to this dilemma is the use of the interstellar probe. Of course, the interstellar probe will only work if extraterrestrial civilizations are very widespread within the galaxy, so that the probes do not need to be sent over very large distances. The solution involves sending out thousands of interstellar probes to the nearest likely stars and arranging to have them get in touch with the local intelligent beings and tell them how to contact us by radio. If their civilization has not yet developed the necessary technology, the probes will wait in orbit for that technology to develop and will then contact the civilization. In the meantime, the probe will radio back to the solar system the relevant physical and chemical data concerning the planetary system which it enters.

The problem of constructing such an interstellar probe lies well beyond our present technological capability. We have not yet developed the technology for building space probes within our own solar system which can be considered reliable beyond 4 to 6 yr of operation. Reliabilities greater than this must be developed if we are even going to send probes to investigate the outer planets in our own solar system.

In the case of a probe sent to another star, it would be extremely wasteful of economics and energy resources to design such a probe to travel faster than 1 percent of the speed of light. One could imagine developing a technology that would permit a faster speed, but the resulting probe would be vastly more expensive, and it must be remembered that these probes to other stars may have to retain reliability for millions or even billions of years. Such are the geological ages over which new biological species evolve. A journey of even 100 light-years at 1 percent of the speed of light is very short compared to these times. It is surely better to devote our technological resources to making sure that such a probe remains reliable for a long time. It must have a remarkably high degree of artificial intelligence of the sort considered by McCarthy in Ch. 5. It must be self-repairing and redundant. Perhaps it should even be reproducing, in the sense argued by Arbib in Ch. 4, so that having reached a given star, it can build a replica of itself which can proceed still farther to other stars and planetary systems.

Man has voyaged to the moon and has found that voyage extremely expensive. The costs will become less with further technological development, and no doubt there will ultimately be more manned visits to the surface of the moon. These visits were all preceded by examination of the moon by space probes.

We are beginning to talk about manned voyages to other planets, possibly Mars, as occurring within this century. Such visits may become feasible, but they would certainly be preceded by many unmanned probes which will determine the details of the environment into which man would be injecting himself.

Will man ever visit the other stars? This is an undertaking vastly different from visiting Mars. The same questions of energy resources which make desirable a speed not much greater than 1 percent of that of light in the case of an interstellar probe also apply to manned space flight. It seems better to spend the extra financial resources in developing cryogenic methods of preservation of the men who must make the long journey, than in attempting to acquire the tremendous energy resources in a rocket that would enable travel to take place at a higher fraction of the velocity of light. Would any man want to spend several centuries in cryogenic suspension on the way to another planetary system, to devote several years to the scientific study of that system, to return over several more centuries to the earth in cryogenic suspension, and finally to report his results to a home civilization that had evolved beyond all recognition? This is doubtful in the context of today's world. Perhaps it will be more feasible in tomorrow's world, but it will always remain vastly cheaper to send a probe, which does not have to return, since it is so much cheaper to radiate back the results of the scientific investigation of another planetary system by radio. It is precisely the same type of argument which one can use to say that we should attempt to see if anyone is sending by radio in the first place, so that we do not even have to go to the expense of sending a probe to that planetary system, but we can ask the inhabitants to tell us about themselves and their environment.

These arguments should be equally applicable in reverse. We should not expect visitors from other planetary systems. Only if vastly new laws of physics are discovered, which will permit such visits to be made cheaply and in a relatively short time, should we expect such visits to our planetary system to occur. It is sometimes thought that unidentified flying objects (UFOs) might be visitors from extraterrestrial civilizations. If so, they have been remarkably elusive, and they must possess that radically new technology.

But we have no rational basis for accepting this interpretation of the UFOs, and the UFOs remain a scientific mystery, which could be purely psychological in their manifestations. In the absence of a contrary demonstration, our only rational and logical procedure in considering extraterrestrial civilizations and how to communicate with them, is to discount the possibility of radically new laws of physics about which we know nothing, and to consider how communication can be brought about using the laws of physics with which we are familiar.

7

Interstellar Probes[1]

R. N. Bracewell   *Stanford University*

Introduction

Much of the stimulating discussion of interstellar communication assumes the use of radio waves for communication. The arguments in favor of the use of radio waves are sound except in a certain case where interstellar probes merit consideration. In this contribution to the discussion of interstellar communication, the case for the use of probes will receive emphasis.

After contact has been made with another civilization, radio is in all cases suitable, and probably optimal, for communication, unless the physics of the future circumvents the serious time delays suffered by electromagnetic waves.

In the precontact phase it is better to consider the situation as a function of d, the distance to the nearest superior community. If d is small, search by radio will in time succeed. At a certain larger value of d, radio search has serious handicaps, and wider attention should be given to the discussion of probes. If d is larger still, further considerations enter as will be seen below, and finally if the nearest superior community is extragalactic, the situation changes again. Table 7.1 summarizes the distinction between the various cases and incorporates values of d characteristic of each case.

The term *superior community* is a key one in this discussion and we must first understand what is meant by it. By superior is not implied any

[1] The material of this chapter is based on a talk I gave at Green Bank in March 1960 after reading Cocconi and Morrison's stimulating paper in *Nature* **184** (1959), 844, and papers by Su-Shu Huang. Further details connected with the idea of dividing the problem into cases in accordance with the distance to the nearest neighbor, with the consequence that longevity is connected with distance and with the use of a probe will be found in *Nature* **186** (1960), 670 and also in A. G. W. Cameron, *Interstellar Communication.* Benjamin, New York, 1963, p. 232 and 243.

Table 7.1

	I	II	III	IV
d, light-years	10	100	1,000	Extragalactic
Search	Radio	Probe	?	Radio
Communicate	Radio	Radio	?	One way only

moral or ethical superiority. A community is superior when it is able to communicate by radio waves, fire rockets into the upper atmosphere and into surrounding space, let off atom bombs, and in general do all the technical things we can do and have been able to do for some time. It might be argued that it is conceivable that a civilization has radio communication but does not know about nuclear reactions. It is conceivable but not likely. Perhaps a civilization has radio and is unable to launch rockets, as indeed was our situation for some 50 yr or more. However, note that the time interval is quite brief, and we can assume that a community, as it learns about physics, passes through the abilities mentioned in fairly rapid sequence. It might take them 100 yr or even a little longer. It is possible that a community might live on a very heavy planet where gravity is so strong that it is very difficult indeed to launch a rocket. Or there may be a planet where all the uranium has sunk out of easy reach; perhaps it was covered by lava that was not rich in heavy elements and therefore it is probable that nuclear fission will elude them for a long time. However, we shall assume that, apart from some fuzziness due to time, everyone knows whether a community is ahead of ours or not, and we can call that a superior community. It can further be argued that there may be communities that are superior to ours that are not interested in launching rockets. It is true that there might be a planet somewhere which is inhabited by very wise men who sit there and think and are very knowledgeable but are quite clear in their minds that they do not wish to communicate with anyone else. This situation is quite possible, and they may know a lot more physics than we do. They may know how to set off atom bombs but may not in fact have done it. Or, they may know the full theory of satellites, just as Newton did, but have decided not to build one. Well, if such a community exists and there is no reason why one should not exist, then they are not the ones with whom we will be dealing. So we can exclude them. They are not among the communities that we are discussing, although they may very well exist.

Case I, Abundant Life

In this discussion we shall split the whole range of possibilities of communicating with other superior communities in the galaxy into three cases. First we shall deal with what we can call the Ozma case. Project Ozma, as you will recall, consisted of turning a radio telescope on two relatively nearby stars Tau Ceti and Epsilon Eridani, which are 10 or 11 light-years away. After a short period of listening with no meaningful message having been received, the project was terminated. No one has repeated the project or suggested a better one. It seems that it is a good thing to do if the nearest stars that we think are likely to be inhabited are in fact inhabited. By a habitable star we mean one which has a zone around it in which the temperature is moderate: it is not boiling, which would sterilize life, and it is not frozen, which would slow life down considerably. In addition to these considerations, stable circular planetary orbits must be possible. Circular planetary orbits would not be possible, for example, around a double star, where a planet could not travel in a circle but would travel in a more complicated orbit depriving it of the billions of years of stable conditions that are probably necessary for life to evolve. So if we eliminate double stars, and stars that are too cool (meaning that the habitable zone for the planet would have to be so close to the star that its thickness would be so small that the probability of there being a planet there would be small), and stars that are too hot (it seems likely that the very hot stars don't have planets), we are left with stars that we can call "likely stars." The fraction of stars that cannot positively be eliminated as above is about 10 percent in the neighborhood of the sun. Of the likely stars, only a fraction will in fact have life, and an even smaller fraction will have a superior community, which has developed from that life.

As we move out from the sun, Epsilon Eridani is the tenth star we meet, but it is the first likely star, since other nearer well-known stars such as Sirius and Proxima Centauri were rejected on various grounds. If the first star we get to that seems to us to be not completely ruled out proves to possess a superior community, would not that be very surprising indeed? Would it mean that every likely star has a superior community? It certainly would imply, barring a fantastic coincidence, that the universe was heavily populated with superior communities, not only heavily populated with life, but with communities that were able to send us radio communication that Project Ozma might have picked up. Thus we can conclude Project Ozma is a worthy but extraordinarily optimistic project to engage in. Project Ozma is the correct experiment to perform if the universe is really crawling with life.

The Basic Graph

From the known star density as a function of position in the galaxy one can work out the number of stars within a sphere of given radius centered on the earth. At first this number increases as the cube of the radius. When the sphere bursts through the top and bottom of the galactic disk, the rate of increase of volume occupied by stars slackens off, but on the other hand, the increasing density of stars towards the galactic center compensates. Finally, at a radius of about 100,000 light-years, all the 10^{11} stars of the galaxy are enclosed in the sphere. The number N_L of likely stars within a given radius rises from 1 at a radius of 10 light-years to 10^{10} at 100,000 light-years. This quantity is shown in Fig. 7.1. The curve can't be claimed to be very accurate, but it can hardly be very wrong either. It is well within a factor of ten and maybe a factor of three.

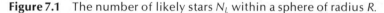

To survey the situation as a function of the density of life, we now construct the graph of Fig. 7.2, in which the quantity d, the distance to the nearest superior community, is our indicator of the range of possible situations. If Project Ozma had succeeded, we would have $d = 10$, and life would be abundant. If $d = 100,000$ light-years (which is about the diameter of the galaxy), it means that superior communities in our galaxy are vanishingly rare. We will discuss $d = 100$ and $d = 1,000$ in turn, after first establishing some quantitative facts.

Let N_c be the number of superior communities in the galaxy. Although we do not know this number, a moment's thought will reveal that we can draw a graph of N_c versus d because each in its way is connected

Figure 7.1   The number of likely stars N_L within a sphere of radius R.

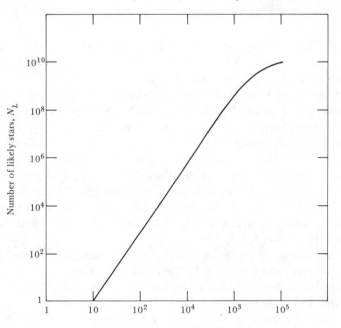

Radius of sphere, R (light years)

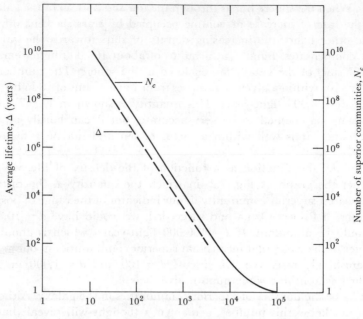

Distance to nearest superior community, d (light years)

Figure 7.2   The total number of superior communities in the galaxy N_C and the average lifetime Δ, as they depend on the distance d to the nearest superior community.

with the density of superior communities and, therefore, there must be a relation between them, which is shown in Fig. 7.2. Of course, the relation is statistical: perhaps $N_c = 1$ and $d = 12$.

Accidents of clustering must certainly occur, and will have important impacts on the chances of contact, helping where there is crowding, and hindering where there is not. The graph in Fig. 7.2 is based on the expected distance to the nearest neighbor when the stars are uniformly, but randomly, distributed in space, but it has to be admitted that galactic structure superimposes features that have not been taken into account. Such uncertainties in d seem unimportant compared with the uncertainty in N_c which, for all we know now, can have any value from 0 to 10^{10}. So it seems that the graph in Fig. 7.2, rough though it may be, can command your assent.

As a numerical example, let us examine the case where the number of superior communities $N_c = 10^7$, and clearly this means that out of 10^{10} likely stars, the frequency of occurrence, p, of superior communities is only one in a thousand ($p = 10^{-3}$). Reading off from Fig. 7.2 we see that $d = 100$ light-years. As a check, let us refer to Fig. 7.1, where it is seen that for the number of likely stars within a radius of 100 light-years we have $N_L = 1,000$. As a second example, we find that when $d = 1,000$ light-years, $N_c = 10^4$, $N_L = 10^6$, and $p = 10^{-6}$.

Case II, Less Abundance

If one in a thousand of the likely stars is the home of a superior community, presumably very many more are the site of life of some kind. Therefore, even this case represents rather common existence of life, and the corresponding distance to the nearest superior community, as read off from Fig. 7.2, namely $d = 100$ light-years, may be felt to be an underestimate. However, as we are examining a number of representative cases in turn, let us consider how to go about making contact when we have to face the likelihood that only one of the thousand nearest likely stars has a superior community.

It is often supposed that the more advanced community will take the initiative and can be counted on to be beaming a powerful radio signal in our direction. Even so, and even if we were listening, there is only one chance in a thousand that we would be listening in their direction.

In the case of the two stars mentioned previously, we knew where to point our telescopes, but now when we go through the full list, we find about 1,000 stars that we could choose from. The problem becomes more complex when we go to the transmitting end to transmit to this other community which is trying to communicate with us, because they also have 1,000 choices of where to point their telescopes, not the same ones, but they are also surrounded by 1,000 possible stars which to them seem not eliminated as possible habitats for life. They cannot tell by telescopic inspection whether the earth is inhabited—they have to make a guess. They really will not know the micro circumstances. For instance, if they had been looking at the earth with the best telescope they had a couple of thousand years ago, it would have looked essentially the same then as it does now, as seen from that distance. And yet, it wouldn't have been worth their while sending radio waves, in fact, even 100 years ago, because there was no one here who could pick up radio. Mind you, there is always the chance that radio is not the final answer and that we still have not qualified scientifically for entering the galactic communications network because we still have not discovered the next thing that waits downstream for us in physics, and it may be that everything in interplanetary communication hinges on this next discovery. We may be a little overconfident in thinking that we now have the necessary tools, as we do not know what the future holds for us in physics.

Thus we are faced with a situation in which they have 1,000 choices to make, and we have 1,000 choices to make so there is only a one in a million chance that we will be listening to them while they are transmitting at us. Now that is pretty tricky. Professor Morrison has an interesting calculation on how they would apportion their time. He says that they might transmit for a whole day in the direction of the earth and then for a whole day in the direction of one of the other choices and so on and so on and so on, and after some years, they would come back and give us another whole day. Now you see what a fragile plan this is. He credits them with enough wisdom that, having transmitted to us for a whole day, when a time had elapsed equal to the time taken for the message to go to the earth and come back, they would turn their big ear back on us and listen to our reply. But suppose that we are a little slow in decoding the message and do not reply immediately, but hold out for a day or two; they might have hung up by the time our message gets back, and if our reply has to take 100 yr to get there, and they just happened to have put the receiver down at the moment our reply came in, it would be very disappointing, but to be expected of such a risky technique for making contact.

In addition to the one in a million factor, which is geometrical, there is the problem that we do not know how to choose which wavelength to listen to. An excellent suggestion was that we should use the 21-cm line because in the general range of wavelengths which seems suitable for interplanetary communication, it is the one wavelength which stands out as being chosen by nature to do its own broadcasting; hydrogen atoms which are a most frequent constituent of the universe give out faint radiation on that wavelength. So they know very well that if we have any scientists here, we know that wavelength and will be tuned in on it. But scientists are not certain that they will use it. Some say twice or half that frequency would be better, so if we were listening on twice the right wavelength, we would miss contact, even though we were doing approximately the right thing. You can also see that there are risks arising from the rotation of the earth; they might spend 12 hr transmitting to us when we are on the other side of the earth. Furthermore, it will have occurred to everyone that this requires political stability for a period of 200 yr; it means that once we have raised the money to do the job, we have to have the same kind of funding carried on for about 200 yr. Now we have not had science funding for very long and we have not had a project with dependable funding for a whole generation, so it is hard to see how we can plan for more than a whole generation. Thus we are confronted with very serious difficulties.

Part of our problem in thinking about this is that we tend to think of communication like a telephone conversation, I say a few words to you, and you say a few words to me, and there is an interchange. But interstellar communication will not be of that kind, quite clearly. There will never be a conversation of that character between two inhabited planets. You will be able to say, "Hello, how are you?", but 200 yr from now it will be a descendant of his, many times removed, who says, "I'm fine, who are you, he was fine." Thus the communication that we are discussing is of an entirely different nature. It is an interaction between two cultures. Two cultures will be able to influence one another, but individuals will never be able to interact. The cultures can interact in a very real way in the sense that we could suddenly find ourselves in possession of the next 100-yr worth of physics and chemistry. That would have a real effect on our community—to suddenly be presented with the discoveries that we would expect to spend the century making. That would be remarkable. It would no doubt have a very strong impact on biology and medicine. If the people in 1870 had suddenly been presented with the next 100 yr of physics and chemistry, it would have made a very great difference to things as they actually developed. So communication, once started, will be on a time scale that is slow by individual standards, but it will nevertheless be very real.

In the preceding discussion we have tried to make the point that if 100 light-years is the distance to the next inhabited community, trying to reach them by radio will present extraordinary difficulties and may be doomed to failure. So we must apply our minds to a procedure that would contain the key element needed to set up communications.

Let us suppose that you sit down to teach someone you do not know to play chess. You could read him the full rulebook, without his saying a word, and at the end he might in principle be able to play. But we do not do it that way. You explain that the rook travels in straight lines, and he questions the direction in which the rook may move. You answer that question, and now he knows how the rook moves, but until you have that little bit of feedback, and there have been a certain number of round-trip loops, he really does not know for certain. You will make a few oversights or omissions that need verification, so there will be a certain number of round trips. You will not be able to do it in one. To start with, if you have never met him before, there will have to be at least one interchange while you decide whether he speaks English or French. And the interstellar case is one where you have not met the people before. So a certain number of conventions have to be established, which require loops. In ordinary conversation we have so many of these feedback loops that we just lose count of them. We throw them in in great redundancy. But if each one took 200 yr, that would make the picture entirely different. It seems that such feedback is extremely effective and in order to have it, we just have to reduce the loop time. Thus we arrive by that line of reasoning at the conclusion that in order to set up the conventions, to set up our wavelengths, to set up a time schedule for transmissions so that the sun or the earth does not get in the way, to agree on a code (this will not be trivial—how are we going to discuss how to discuss what language we will use?) will require these round trips, and they must be reduced to essentially zero time so that a large number of them can come well within the lifetime of one individual who can handle the whole thing. All those round trips must take place with political stability and also within the stability or permanence of a single individual.

The way to set up the conventions is to have a probe which would be launched from the one planet into the vicinity of the other planet. When that probe arrives, we have the possibility of communicating with it on a rapid give-and-take basis, and we can set up all the conventions needed so that the discovery or contact phase is handled in that manner. Then, when we have set up all the arrangements, we can go right ahead and communicate back to home base with a prearranged system. That of course will take us 100 yr—there is no way to avoid that. Nevertheless, a stream of information will then begin. We will start streaming information towards them, and since we are going to have to wait 200 yr for a reply, our initial statement had better be a very long profound statement, something taking about 100 yr to say. Now that statement will consist of essentially our whole culture. We can transmit the whole of the *Encyclopaedia Britannica, Webster's Complete Dictionary* with illustrations, and all our music. Our art might prove baffling to them, but they would probably be interested in it, and we would have to microfilm the contents of all our museums. In fact, it would be difficult to find something that would take 100 yr for us to say, unless we sent them telephone directories of all cities, and threw in old sports results for good measure. Then we would get something back from them, and what would happen then is another story.

That is the proposition. When their probe gets here, there are various conceivable possibilities, but this author feels that there has to be a probe. The probe could simply be one aimed at detecting whether there was life here. It could arrive here, look around, see signs of life, and then radio back to the home base that this was the place to shoot for. If that is the situation, then that message could already be on the way back. The probe could have arrived, it could have been circling around in our solar system, heard our first radio communication, and sent word back. So the message could be on its way back now. The difficulty with sending such a probe is that most of the planets that such a probe visited would not be inhabited. Therefore their probe would probably be equipped for exploration not exclusively concerned with the finding of life. It might count the planets, measure their general parameters, size and shape, and do all sorts of space science and report back those basic statistics. That would be of some interest—it would be an extra entry in their catalog. But even that is fairly dry, so they will probably include a beacon for the purpose of attracting our attention. It would be a shame if it had been here, sent back word, and was now lying dead out there near Jupiter somewhere to be photographed by the Pioneer spacecraft, with its lifeless antennas all sticking out. It is not likely that they would do that. They will send one here that will attract our attention, and when it has attracted our attention, then it will talk to us. So there are at least three phases in this whole procedure. It has to attract our attention, convey the identity of its home base, and set up codes and schedules for communicating with the parent star.

This phase completely circumvents the risk that we are not listening on the day they have allocated to us, since the probe will remain in action once it arrives here. The part of the budget of their Space Administration devoted to interstellar contact is spent not on power for hit or miss transmissions but on launching one probe at a time at a rate in accordance with the ruling economic and political conditions. As each probe reaches its destination, it fires its one remaining rocket and goes into orbit around the sun at the distance where the temperature is right, deriving its power from then on from the sun. It could of course carry its own nuclear fuel, but since it might have to remain on the alert for thousands of years, it would be wiser to make use of solar energy. That might seem like a long time to wait patiently, but the same patience would be required if radio transmissions were beamed directly from the distant base. There is a big difference, however, between the ongoing running costs in that case and the one-shot expenditure in the case of the probe.

On arrival, the probe will turn on its radio receiver and listen for manifestations of intelligent life. The presence of radio signals will inform it beyond doubt that communicating entities are about. No doubt it is feasible to design an elaborate probe that can locate and descend to a planetary surface, but that would not represent an optimum search strategy. On the contrary, the more modest the probe the more planetary systems can be explored. The rather simple orbiting radio receiver is an effective means of inducing those communities that have reached the point of communicating, to report in.

How they might attract our attention requires some thought. The idea of powerful bursts of radio or optical emission, perhaps separated by intervals to allow even larger peak power, suggests itself, but some precautions are needed to guarantee against failure on our part. Even the emission of a very strange program would not necessarily attract our attention when you consider the psychology of people tuning radios.

The following is a proposal for a radio transmission that automatically guarantees that the frequency is in use on the earth, that ionospheric or other effects will not prevent its reaching the earth, and that when the signal reaches the earth there will be a receiver tuned to that frequency at the time. The plan is simple, it requires little power, and it is free from problems connected with the ionosphere, local interference, and time schedules. You might be sitting, turning the dial of your receiver, looking for some particular program, but you wouldn't stop to listen to things that you are not interested in. Perhaps you are trying to pick up the news from Moscow, and you get this faint rumbling sound which is the first message to man from interstellar space; you listen to it carefully and as soon as you can distinguish that it's not what you want, you just tune away from it. Yet it could be the most conspicuous thing. It could be a voice saying, "I am the man from outer space," and you would immediately tune away from it, because it's not what you want. To assure that its transmission would be noticed, the probe would listen to what it could hear, and whenever it hears a broadcast, it transmits on exactly that wavelength. This automatically guarantees that someone is listening, because no one transmits if nobody is listening. It knows there is a receiver tuned to that wavelength, so it gives out on that wavelength. What should it give out? If it gives out something that you don't want, this may just turn you off, you will go away. We propose that it would transmit back what it heard; thus if there was speech coming out it would transmit the speech back. We would hear the program that we wished to hear, followed by an echo, with a time delay corresponding to the time taken for the signal to go out to the probe and back, which would be of the order of minutes. Now that's something that would be very conspicuous indeed. We know that this is so because on the rare occasions when people do hear long-delay echoes, they always notice them. The people who listen to long-distance radio are very familiar with the echo produced by passage of a radio wave completely around the earth which takes 1/7 sec, and they get used to that time delay which is quite recognizable as such. But when you hear the announcer say, "Well, that's the end of the news, goodnight," and then 5 min later, just as you are about to get up and switch the receiver off, a voice repeats, "That's the end of the news, goodnight," you will certainly notice it. That is why it seems that it would be a good strategy for them to do that very simple thing—just install a receiver, an amplifier, and a transmitter.

How would we tell it that we have heard it? Until we do that, we will not hear the message that it contains. There would be no point in its spilling its program as soon as it arrived here; first it has to attract our attention, and then we have to let it know that we have heard it before it will come out with the good news that it has.

It seems that we can build up a complete system of communication on the very simple idea of repetition that has already been introduced. All we would need to do after the first excitement of noticing that the echoes were reproducible, would be to set ourselves up with nothing more than facilities for repeating back to it. How conspicuous to it that would be! In the early twentieth century it would have begun to notice our radio emission, and after years of listening to our programs and playing them back endlessly, all of a sudden one day, it would get played back at itself. It would be preprogrammed to notice that, and it would immediately deduce the time delay, and hence how far away we were, and it would know that it was in communication with us. To verify this, it would repeat back to us again, and we would again repeat back.

This procedure is similar to the one used by the Count of Monte Cristo. You remember he was imprisoned in the Chateau d'If which had extremely thick walls, and the next dungeon was thirty feet of stone away, but he got into communication with the prisoner there by tapping on the stone. The first time he tapped, the man on the other side heard it but did not do anything, but after days of hearing these taps it occurred to the other prisoner to tap back, so he tapped back. Now that tells the first man that the second man has heard him, but it does not tell the second man that the first man has heard the reply. What is the solution? When you get that exciting first return, then the next thing to do is set up a code so you send out two taps, and then when two taps come back, you and he both know that you are communicating. Now you are on the way to building up a higher form of communication by somehow agreeing on a Morse-type code. But how do you teach the man on the other side of the wall Morse code when neither you nor he knows Morse code, though you know that such a code exists.

The problem is quite fascinating and could be the basis of a game in which you exchange notes with a friend on the understanding that you would use no known code and that communication was to be established starting from absolutely nothing. It would be quite challenging.

There would be some essential preliminaries in the dungeon problem such as determining when the other prisoner sleeps or is otherwise preoccupied, and clearly the acknowledging tap permits this. If one had to tap out signals in Morse without knowing whether anyone was listening, the chances of being understood would be hopeless. The closed loop makes the essential difference.

Similar preliminary essentials confront the probe. Soon after the first contact, it is likely to set below the discoverer's horizon, and then to our great disappointment we will realize that it is no longer there. It is very important for the probe to discover at a very early stage precisely what schedule we plan to keep. It can do this by monotonously going through repetitions in order to avoid loss of part of its message. It can determine the sensitivity of our equipment by turning its volume down, and as it turns its volume down, it gets to the point where we no longer respond. Then it brings the volume back up until we do respond. It can also tell the band over which we can tune. Since we are moving relative to it, it knows we can tune our receiver to compensate for Doppler shift, so it deliberately goes off frequency a bit, and then of course we discover what it is doing and repeat back, and it goes a little farther, and this way it will discover that when it gets down to the penetration frequency of the ionosphere, we give up because we can not get through. Then it can explore in the other direction and find that the oxygen and water vapor cut us off at another wavelength. It would like to know the information rate that we can accept, because if we are not capable of recording very rapidly, there is no point in its playing its message too fast. It can find out all these things and many other details just on the basis of knocking on the wall and listening for returns. Quite a sophisticated picture of our capacity for communication can be built up by this simple preprogrammable technique.

Preliminaries of the simple kind mentioned above may determine whether the whole project succeeds or fails. To give an idea of the contingencies for which the probe must be prepared, just imagine a planet where afternoon thunderstorms during the summer are of such intensity that all radio communication is overwhelmed by static. Or suppose that ionospheric storms causing radio blackouts come with greater frequency than on the earth and that the solar rotation period of 27 days and the solar cycle of 11 yr, which dominate such things, were different. Outbursts of solar radio noise are not an embarrassment to radio communication on earth but might well be in another planetary system. Thus before launching its message, the probe must satisfy itself of many preliminaries to avoid sowing its seed fruitlessly.

Methods for deciphering messages arriving direct by radio from alien planets have been discussed by others, and it is generally considered to be feasible, so we need not dwell upon the matter other than to remark that the situation is improved when the possibility of feedback is present. Without feedback, massive redundancy is required to guard against unforeseeable losses of unknown duration with a consequent heavy penalty on the chances of success in making a successful transfer.

Case III, The Nearest Is Remote

Despite the difficulty of doing what has already been proposed, nevertheless let us now suppose that it is 1,000 light-years to the nearest superior community. In this case we can do a further calculation which is really quite fascinating. Table 7.2 summarizes some earlier calculations for reference and includes similar results for $d = 1,000$ light-years where we see that the number of superior communities in the galaxy $N_c = 10^4$, and their frequency of occurrence among likely stars is given by $p = 10^{-6}$. If among 1 million likely stars, only one has a superior community, then under a condition that we may refer to as secular equilibrium, the average

Table 7.2

d, light-years	N_L	N_c	p	Δ, yr
100	10^3	10^7	10^{-3}	5×10^6
1,000	10^6	10^4	10^{-6}	5×10^3
2,000	10^7	10^3	10^{-7}	500

duration Δ of a superior community will be p times the time taken for a likely star to produce a superior community. Adopting 5×10^9 yr for the latter, we find that $\Delta = 5,000$ years. It may seem surprising at first sight that given the distance to the nearest superior community, one can then estimate how long it has before collapsing, even if only on some average basis. But suppose that the duration of a superior community is 5×10^6 yr, that is, 10^{-3} of the preceding evolutionary development. Then it is very hard to see why the frequency of occurrence of superior communities should not be one in a thousand, with the consequence that the nearest neighbor, on the average, could not be as far away as 1,000 light-years.

This type of calculation requires elaboration to take into account the changing parameters as the beginning of the universe recedes into time. Perhaps there are existing relics of a chain of galactic communication that was established long ago when distances were smaller. But there has been ample time for the birth and death of stars and their civilizations, if any, as witnessed by the fact that our own sun is a second- or third-generation star condensed from chemical elements that themselves were brewed in earlier stars, long since exploded. Consequently, the statistical calculation of duration, though crude, has to be faced.

If $d = 2,000$ light-years and $\Delta = 500$ yr, it appears that Case III confronts us with a situation where the longevity of civilizations is not sufficient to permit both of the participating civilizations to know that contact has been made.

We should not despair entirely at this, because even a one-way flow of information would be significant, both for the recipient of the alien culture and for the sender, whose interest would lie in the preservation of its tradition, though not of its population. Nevertheless, the technical feat of transfer is even more difficult to imagine.

Of course, there can be all kinds of departures from an average, so even if N_c is only 10^4, some of them may have solved the problem of how to survive longer than the average. They may have gained control of the circumstances that lead to short average lifetimes, and for all we know, by accident of proximity, some of them may be in communication with one another and have spread their secret. But it will not be easy for us to be inducted unless by a further accident.

Case IV, Extragalactic

If the nearest superior community is not in our galaxy at all, it is quite clear that we must point at the Andromeda Nebula and explore carefully for a radio signal. We know where to point, and they know where to point so the geometrical difficulty is eased. But how do we know there is no other superior community in our galaxy? We do not, but after a reasonable interval of looking for nearby life, we might as well try elsewhere. The message has to be one of extreme redundancy so that one can tune in at any time and still get something; it is a strictly unidirectional flow. Would it even be worth answering? Of course, if we first make contact within our own galaxy, we will get the answer to what, if anything, is streaming to us from Andromeda, as well as a partial story on the lines of communication that are already buzzing within our galaxy, so we should direct our efforts at local contact first.

Conclusion

We conclude that $d = 10$ implies improbably high density of life, $d = 1,000$ requires improbable durability, and that $d = 100$ is the case we have to face seriously, even though here also a remarkably high density of life of all types is implied.

Careful attention should be given to the method of contact by probe, but, because of the times and distances involved, we will have to be patient. Our contact, when it comes, will not be the first of its kind and therefore we will be brought into touch with a chain of communities already in communication with each other who know quite well how to go about doing it and who are not as impatient as we are. They know they only succeed every few centuries (?) in adding another link, and they are not in any hurry, whereas we, of course, wish it would happen sooner. They may have tried us 1,000 yr ago with a probe that has now run down and be planning to try us again 1,000 yr from now.

Even when contact comes, the flow of information may for a long time be unidirectional—one might say unidirectional both ways. We will receive the impact of the quantity of information that the probe brings with it, and long after that their home planet will begin to receive a stream of material from us, but it will be even longer before interaction reaches the point of our receiving answers to questions. Or perhaps this underestimates the capacity of a probe—surely biology teaches us that the capacity of a human individual can be contained in a space the size of a human head. Therefore, why should not a substantial reference library be compressible into a further small volume. Thus, when our probe comes, we can prepare for a major cultural impact which will be greater than if our first contact is by direct radio.

The Message

Bracewell has told us of some of the difficulties of establishing communication with extraterrestrial civilizations. In the following chapter, Frank Drake investigates some of these difficulties further, including our present technological capability of sending and receiving signals, and the problem of what that signal content should be like.

One of the problems raised by Drake is the problem of the message content of an interstellar cosmogram. Drake tells us that one of the highest information content types of messages would be a two-dimensional picture. The information content of such a picture is very high, only if we can utilize our own cultural and technical background to draw inferences from simple representations within the picture.

Drake is probably right that this is the best way for a message to begin. This is probably the best way to define symbols that will be used in one-dimensional language communication later on in the message. The representational technique of laying out a message with the number of meaningful cells being the product of two prime numbers could be generalized to the third dimension if this was desirable for some special purposes.

However, it must be remembered that the time interval required between sending a message and receiving an answer to specific information content within that message is probably many years, tens of years, or centuries. Most of the information content in the world literature lies not in two-dimensional representations of data, but rather in one-dimensional representations of data called words and sentences. Thus the type of picture technique described is probably a good teaching tool to demonstrate to us how to receive the remainder of a message, which may be centuries long in transmission, and which may have further pictures as illustrative material throughout. However, if the sender were not sure that his message was being received, he would have to repeat his tutorial material frequently throughout his message, if he wanted to make sure that someone coming "on line" in the middle would be able to understand it.

8

Methods of Communication: Message Content; Search Strategy: Interstellar Travel

Frank Drake   *Cornell University*

In this account of the concepts and ideas related to radio communication with or detection of other extraterrestrial civilizations we shall first examine the basic concepts upon which the search for extraterrestrial signals is built and then discuss the matter of radio detection and communication.

The problem of detecting other civilizations can be reduced to two questions. The first question concerns the probable distance to the nearest detectable or communicative civilizations. We must solve that question before we can turn to the second question, the choice of the most successful method of detecting other civilizations. Obviously, we must answer the first question before we can attempt to answer the second.

The distance to the nearest detectable civilizations can be determined by first discovering the number of communicative civilizations in the galaxy. Once we have that number, knowing the distribution and density of stars in space, we can quickly determine the distance to the nearest civilizations. So we need to know the solution to an equation for the number N_l of communicative civilizations in space. This number is proportional first to the rate of star formation, R_*. The rate at which possible abodes of life are created in the galaxy is proportional to R_* times the fraction of such new stars which have planetary systems, f_p. Thus $R_* f_p$ will be the rate of production of planetary systems per year, and these numbers are fairly well known. The astronomical evidence shows that R_* is about one star per year in the galaxy; this has been accurately determined from the theories of stellar evolution and the simple counts of different types of stars in the sky. The fraction of newborn stars which possess planetary systems is less well known and has not been observed directly. There is some inconclusive evidence for other planetary systems. That evidence, plus the statistics of binary stellar systems and the theories of stellar formation, indicates that f_p is probably of the order of 0.5. About one-half of a new planetary system is produced per year in the galaxy.

Each of these new planetary systems will have some number of planets, n_e, which are in the ecosphere, the sphere of life which surrounds a star. Using either our own system as an example or based on the theories of planetary formation, we conclude that n_e is of the order of 2. In the solar system, for example, in addition to the earth, Mars and Jupiter are possible abodes of life. The product $R_* f_p n_e$ indicates that a new planet capable of supporting life is created each year.

Some fraction of these planets will actually give rise to life. Here, of course, is a problem for the biochemists. Given a planet like earth, what is the probability of life arising? As has been discussed in previous chapters, that probability, f_l, is extremely high, that is, life is not something that occurs by accident—it is essentially inevitable, given a planet with a chemistry like that of the earth. The number for f_l is about 1. Thus, we have a product $R_* f_p n_e f_l$ which states that there is one new system of life created each year in our galaxy

Some fraction of these planets supporting life, f_c, will give rise to a communicative civilization. Here we are getting to a factor which is much more controversial. It concerns the problem of whether the evolution of life, once started on a planet, inevitably leads to an intelligent technical species, or whether the path of evolution can dead end before it reaches a species which is communicative or intelligent. The number for f_c may well be about 1, but this is questionable; we will return to this problem below. Assuming that intelligence evolves, and that $f_c \approx 1$, we have that the rate of production of technical or communicative civilizations in the galaxy $R_* f_p n_e f_l f_c$ is of the order of 1. In other words there is about one new intelligent, technical civilization in the galaxy per year. We now have a picture of the galaxy and the life in it as an evolving situation. New civilizations are rising at the rate of about one per year. We wonder how many there are at any given time. If they all were to last indefinitely, in the 10^{10} yr the galaxy has existed, there would at present be billions of civilizations. If that were the case, it is surprising that we have not already discovered intelligent life.

The absence of the detection of manifestations of life suggests that the longevity of these technical civilizations is limited; therefore, we must multiply by mean lifetime or longevity L. When we do so, this product becomes the equilibrium number of communicative civilizations in the galaxy: $N = R_* f_p n_e f_l f_c L$.

The longevity is a very dubious factor because to know L, we may in fact have to detect life in space to predict it, and thus to learn how easy it is to detect life. The factors that limit the longevity are obvious occurrences such as nuclear war, the capability for which is almost certainly developed simultaneously with the development of the means of distant communication. However, a more serious and more common limit to the longevity is increased technical prowess. To detect other civilizations, energy is required—power which they waste which comes to us unused by them. It can be expected that as civilizations become more developed, they will conserve power more and more and develop those systems which will not waste power. If they do so, they will become undetectable. So indeed the limit to longevity may not be nuclear destruction, or plagues, or cosmic accidents like cometary collisions, but rather increased sophistication. The mean of the guesses for longevity is of the order of 10,000 yr, which is an estimate based on very poor evidence. In fact, we have been a communicative civilization only 50 yr or so. So here is a very major extrapolation.

To estimate the likelihood of the evolution of intelligence the question of the convergence or divergence, both chemical and biological, of biochemical and biological development must be solved. Many, particularly anthropologists, believe that the very complicated fossil record of the earth indicates that the evolution of life has been an extremely random process in which any endpoint might have been reached. Indeed, they believe it is only a very rare fluke that we ever did develop an intelligent species. They believe that the fossil record shows that the evolution of life is very much like a maze. You may take many paths and perhaps come out a large number of doors. Based on that belief is the prediction that perhaps intelligent life is rare among the planets that actually bear life.

On the other hand, if we consider the history of the earth, it appears that some aspects of evolution have always progressed towards certain ends. One of these ends, the one that is most relevant to us, is intelligence. From examination of the fossil records, we find that the one characteristic which has always improved with time is intelligence. Other characteristics have come and gone, but intelligence has always improved. This suggests that despite the complexity of the fossil record and its indication that the development of life is very haphazard, there is an underlying current which leads to one inevitable end—intelligence.

In the terms of the paleontologist the problem is one of convergence or divergence. By a study of certain mazes it is demonstrated that the complexity of the fossil record of life on earth may be misleading and, indeed, convergence rather than divergence may be the rule in biological evolution. Perhaps we will learn the answer from our own studies of the earth, or perhaps we will only find the true answer to this when we have studied life elsewhere. In any case, we like to think, perhaps erroneously, that f_c is 1.

There is one other factor that has a bearing on this discussion. That is, in the history of the earth there have been at times several intelligent species simultaneously existing on the earth. This was true in the case of primitive man. What has happened is that in the land animals, the dominant species, the one that is most intelligent, has annihilated all the close competitors. Again we are misled. We see one intelligent species and we think that it is a small productivity of intelligent species. There have been many, but those that were slightly inferior have been destroyed. This is not the case in the ocean, however. There one finds all brain sizes, all intelligences, up to the most intelligent—the dolphin and the whale. However, in land animals, there is a great gap in brain size between that of *homo sapiens* and that of the next most intelligent creature, the chimpanzee. In the near past the intermediate brain sizes did exist but they were all destroyed by our ancestors.

We are led to a guess at a number of communicative civilizations which is of the order 10,000, that is, 1 in 10,000,000 stars. This means that if we consider the geometry or arrangement of the galaxy, the distance to the nearest intelligent detectable civilization d is about 1,000 light-years. Since this distance varies only as the cube root of N, we can be far wrong in our estimate of N with little harm. For instance, if L is only 10 yr, which we think is too short, d is then 10,000 light-years. On the other hand, if L is a million yr, d becomes 100 yr. We can have a considerable inaccuracy in this equation, and $d = 1,000$ light-years is not too far wrong. Thus 1,000 light-years, despite the fact that it is based on some very uncertain factors, is not too bad a number to use for an estimate of the distance to the nearest detectable civilizations.

In Ch. 7 Bracewell pointed out that this distance is a very serious problem because nothing travels faster than the speed of light. For interchange between civilizations, we expect communication times (conversation times) of many thousands of years, which is very discouraging. We would like things to happen faster than that, and Bracewell argues on that basis that we might expect the use of unusual techniques, such as messenger probes, which would allow communication, transfer of information, to occur much more rapidly. However, that procedure is a very expensive one which requires the launching of many very expensive probes without the certainty that any of them will succeed. In view of this, we need to try other more inexpensive methods to achieve a useful interstellar communication. The problem is to discover an economical way to detect and communicate over this distance.

We need to determine criteria by which we can judge whether a system of detection or communication is a reasonable one. At present there are only two theorems which are almost certainly correct and very useful in determining how we should proceed. The first is the theorem of contemporary mediocrity, which simply means that we must not assume that our technology is terribly advanced and that what we do best is what other civilizations do best. That, of course, is a fairly obvious theorem. We are a young civilization; we should expect that other civilizations have much more advanced technologies so that things we find difficult may not be difficult to other civilizations. In other words we cannot use contemporary technology as a guide in choosing a system of interstellar communication. Rather, we must depend entirely on limitations which are inherent to the cosmos and to all civilizations, and these are the laws of physics and the arrangement of the universe. We must use only systems which are bounded by those things and ignore any implications of contemporary technology.

The second theorem at first glance appears anthropomorphic—peculiar to our civilization—although it is not. That is the theorem that economy is practiced universally, that is, that civilizations everywhere will do those things which are least expensive. They, like us, will use procedures which minimize the needs for personnel, materials, and energy to achieve their ends. The reason other civilizations will act as we do is that all people live on round planets, and round planets have a limited surface area, limited food supplies, and limited energy sources, and this has been in fact the basic cause for evolution in the first place. The competition for limited resources is what leads to improved species. Thus, in fact, the limitations of planetary size and energy resources not only give rise to the intelligent civilizations in the first place, but they automatically inbreed into all the creatures a concept of economy.

Thus we have these two theorems to work with: (1) economy, and (2) contemporary mediocrity. It is very likely that these principles cannot lead to a unique answer about the best way to conduct our search, but they can suggest the most probable ways. We cannot be sure that the most probable way is the one which will work first, because perhaps other civilizations use methods which are not quite what we would think optimum because of some factor in their civilization which leads to other things being optimum for them. We think radio waves are a very probable way to succeed, but this does not mean that this is the only way. There is some small chance, for instance, that spacecraft might be used for interstellar communication. Keeping that in mind, we then ask what is the most economical way to communicate across space?

The use of rockets is an example of a way that is not economical and which is in consonance with the theorem of contemporary mediocrity. They are fine for traveling around the solar system at relatively low speeds, but if we are going to move at light velocities, which is necessary in order to communicate over interstellar distances, the energy requirements become enormous. The weight of a rocket to go 99 percent the speed of light from here to Alpha Centauri, land there, take off again, and come back to earth is of the order of 10^{11} tons. It has a payload of only about 1,000 lb, which is very trivial. This tremendous initial takeoff mass is required because of relativistic effects. As the speed of light is approached, things appear heavier in a way, and this requires more fuel, which itself appears heavier, and so there is a vicious cycle causing relativistic rockets to be extremely impractical. Not only are they massive and expensive, but there are related problems with such a rocket. It uses the conversion of hydrogen into helium, the nuclear-fusion process, for its propulsion source. With that propulsion source, if the rocket takes off from the earth, one hemisphere of the earth will be incinerated by the exhaust of the rocket. Thus rockets are very uneconomical. This, of course is a disadvantage of the interstellar messenger probe discussed in Ch. 7.

An example of an economical method of communicating is the use of radio telescopes. Using a 150-ft antenna with a similar antenna 10 light-years away, we could send a 60-word telegram for a total energy expenditure of about 1 dollar. The use of rockets seems uneconomical when compared with the ability of radio waves to transmit useful information at very low cost; we see immediately that there is enormous economy in the use of electromagnetic radiation. Not only is information transmitted, but it is done at the speed of light, the best possible speed. The conclusion reached from these considerations is that what we will send across space is information, not objects. Objects cost a great deal to send; it is far cheaper to send the instructions for building the object than to send the object itself.

This argues very strongly that our interstellar contacts will occur by means of electromagnetic radiation and not by means of rocketry or solid objects. (It is also one of the prime arguments against the extraterrestrial explanations of UFOs.) We need to determine in which frequencies to conduct our searches or communication with the highest probability of success. To choose frequencies that are better than others in the electromagnetic spectrum, we again turn to the concept of economy, and say that those frequencies will be used which are most economical. This hypothesis may not be correct, but it is very useful and possibly the most powerful one that can be found.

One of the oldest hypotheses is to pick those frequencies that are in a way magic numbers, frequencies which have a universal significance such as the frequency of the hydrogen atom, the basic building block of the universe. Now, however, there are so many magic numbers, so many line frequencies found in the radio spectrum that this hypothesis, which was once so attractive, has lost much of its appeal, since no frequency is overpoweringly significant. For example, strong arguments can be made for using water-vapor lines since our life is based on water as well as on hydrogen and many other things. Thus the concept of magic frequencies is now very weak. However, it may be possible to tie it, as we will see, to the idea of maximum economy.

We assume that the frequency of maximum economy will be used by other civilizations in their attempts to communicate across space. (Note that we are assuming that they are actively trying to communicate. Later, we will discuss the case where they are not trying to communicate, and we will thus not have their help.)

In an electromagnetic communication link there is some antenna with a transmitter of power P_t which would focus our radiation into some beam with gain G_t which goes across space—some distance R—where it is again collected by some energy collector of collecting area A_c. The beam is then delivered to some kind of receiver—possibly a photoelectric tube if this is an optical radiation, or a radio receiver if it is radio waves. Let us call it a receiver which has some minimal detectable power P_d. Now, the power received is of course well known and is given by the equation.

$$P_r = \frac{P_t G_t A_c}{4\pi R^2} .$$

Then the range of maximum detectability is given by

$$R^2 = \frac{P_t G_t A_c}{4\pi P_d} ,$$

which is formed by changing the first equation around somewhat.

We will attempt to show that the limiting factor on the range is nothing more than the minimum detectable power of our energy detectors by demonstrating that the factors P_t, G_t, and A_c in fact do not lead to an optimum frequency. It appears that this method uses our contemporary technology to define our ability to achieve various values of these factors, in violation of the law of mediocrity, but this is not true, because the values achieved are not a reflection of our own technological ability, but indeed of basic physics.

Figure 8.1 shows a plot of these various factors as they existed a few years ago. In the curve A_c (note that the scales are logarithmic) at 100 megacycles there is a very large collecting area. This is of the order of the collecting area of the Arecibo antenna. There are telescopes of the order of the 140-ft at cm wavelengths, and at optical frequencies we have the 200-in. telescope at Mt. Palomar. Note the trend downward which means that given a certain amount of money (and typically the same amount of money is available at all wavelengths), one can build only smaller collecting areas at the shorter wavelengths. This is not so much a result of our own technological prowess, but rather of the fact that at the shorter wavelengths, one must achieve higher tolerances and mechanical stabilities, and these in turn are controlled by the rigidity of materials, which in turn is controlled by the binding forces between molecules and atoms, so indeed the size of the collecting area is controlled by the physics of the universe. Thus this curve, which at first glance only reflects our technology, in fact reflects fairly accurately the limitations forced upon us by the laws of physics. We will expect the same curves in other civilizations; the difference will be that the curve will move up or down. If more money and more resources are available, the curve may move up, but its slope will not change. Note that the G_t curve is nothing more than A_c/λ^2, where λ is the wavelength, and so it too is governed in the same way. Again, we will see the same curve in other

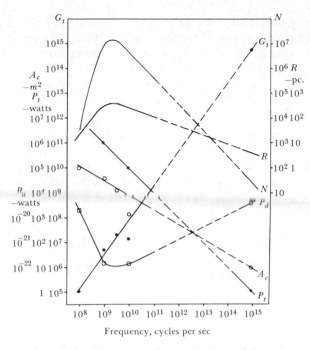

Frequency, cycles per sec

Figure 8.1 A plot of the factors P_t, G_t, A_c, R, N, and P_d.

civilizations with only a change in the vertical level, not in the slope. Similar considerations apply to the transmitter power curve. To generate power at a certain frequency with high efficiency, one must put an energy source in a tuned resonator, and as one goes to the smaller wavelengths, the resonators become smaller, and therefore the power outputs are limited. This is because the maximum field-strength possible in these cavities is limited again by the binding forces in the material. Therefore, the curve for P_t again does not really reflect our technological ability but again reflects basic physical laws. Again, in other civilizations we would expect to see a curve of the same slope but located farther up or down, depending on their stage of development.

The product $P_tG_tA_c$ on a logarithmic plot is obtained by simply adding the three curves P_t, A_t, and G_t. Note that if those curves are added together, we get essentially a horizontal straight line. That's the point—those coefficients, those particular parameters taken together, do not lead to a preferred frequency. This leaves P_d, a curve which is not a straight line but which has a minimum. Now a minimum in P_d, because it is in the denominator, leads to a maximum in range so that P_d leads us to an optimum frequency after all. That optimum frequency is controlled entirely by the minimum detectable power.

We can, in fact, write the minimum detectable power independent of the state of terrestrial technology in a way which applies to all civilizations. This equation is one of the few that holds true all over the universe in the same way. The minimum detectable power (and we shall write it as a proportionality using the radio engineers' terminology) is proportional to the noise which is introduced into our receiving system by the universe itself (we call that the cosmic noise T_c), plus a second source of noise whose significance was first pointed out by Dr. Bernard Oliver. This is the quantum noise T_y, which is associated with all electromagnetic radiation, and which is normally unimportant at radio wavelengths but becomes the dominant source of noise at optical wavelengths. Now these two noise temperatures can be expressed in the following way. There is a brightness temperature for the sky, which is the brightness a black body would have to have to give the radio radiation coming from a certain point in the sky. We will write it as a function of α and δ, the right ascension and declination position in the sky, $T_c(\alpha,\delta)$, and it has a spectrum which is typically a power law, $\nu^{-\gamma}$, where γ is also a function of position in the sky. Thus the cosmic noise is given by this expression and can be measured very precisely. Both $T(\alpha,\delta)$ and $\gamma(\alpha,\delta)$ for every point in the sky have been measured by radio astronomers. The quantum noise is very simply $h\nu/\kappa$ as shown some years ago by Dr. Oliver. We can solve for the frequency of minimum detectable power which is also the frequency of minimum cost, since typically an erg of energy costs about the same to generate at one frequency as another, by simply differentiating $T_s = h\nu/t + T(\alpha,\delta)\nu^{-\gamma}$ and setting equal to zero. We then get the equation which is true throughout the universe. Now ν_0 is the frequency of maximum economy or range and is given by the following equation

$$\nu_0 = \left[\frac{kT(\alpha,\delta)\gamma(\alpha,\delta)}{h} \right]^{\frac{1}{1+\gamma(\alpha,\delta)}}$$

where k is Boltzmann's constant, and h is Plank's constant, so there are two atomic constants. Then we have an exponent which is somewhat peculiar. This leads us to the frequency of maximum economy, and as you can see, it fulfills our requirement: it has nothing to do with terrestrial technology or even knowledge. It involves only the arrangement of the galaxy, whose cosmic noise gives rise to $T(\alpha,\delta)$ and $\gamma(\alpha,\delta)$, and two fundamental atomic constants which are constants of the universe.

Now you will note that because of the presence of $T(\alpha,\delta)$ and $\gamma(\alpha,\delta)$ this frequency of maximum economy is different at different points in the sky. So when we consider different points in the sky, we will arrive at different frequencies. If we take the range of those values which exist, we get as the wavelength of maximum economy a wavelength in the range of 3 to 8 cm. If we wanted to consider a target star in any given direction, we could compute precisely the wavelength which is optimum for that direction. This is now the wavelength at which we can see the farthest. It is also the wavelength at which the transmission of energy information is the cheapest.

Note that in Fig. 8.1 at the lowest frequencies there is a steeply rising curve which is simply the curve of the galactic radio noise, in essence, jamming the signal at the lowest frequency. At the higher frequencies, the rise in the curve is a result of the quantum noise.

Figure 8.2 shows the range and number of stars that are achievable with existing radio telescopes on earth at various frequencies. The frequencies that are best are ones that we can already produce and which penetrate the atmosphere of the earth. Thus we are not in the difficulty of having to go to space or to the back side of the moon to make use of this particular hypothesis. The range one reaches is of the order of 1,200 light years. At the optimum frequency, the number of stars one may reach is as great as 10^7. Not only do we have an optimum frequency for electromagnetic detection and communication, but it is also at a frequency we are good at producing and, best of all, even our existing equip-

Figure 8.2   The range and number of stars detectable with existing radio telescopes on earth.

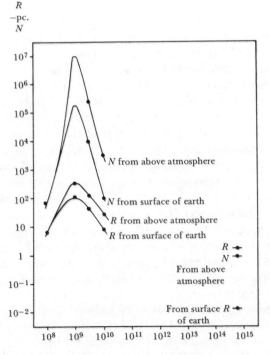

ment can reach the number of stars we must reach to succeed, 10 million. Thus we have entered an era in just the last few years when man for the first time has had the technological ability to detect other civilizations.

Now, we still have to search 10 million stars, so the task is not easy by any means. We have narrowed the band of frequencies that must be searched, but there are still a lot of stars to look at. One might, by the way, combine this result with the magic-number hypothesis by perhaps picking the magic number which is closest to the optimum frequency and using that as the prime candidate for a search. If that turned out to be the spectral line of formaldehyde, for example, you might give that frequency range a special emphasis in any search.

One of the first attempts to listen to signals from outer space called Project Ozma was conducted with the 85-ft telescope at the National Radio Astronomy Observatory. It was used at one of the magic-number frequencies, 1,420 megacycles, to search the two nearest sunlike stars, Epsilon Eridani and Tau Ceti. Although this search did not produce results, only two stars were examined, and as we have seen, we really need to look at 10 million before there is a good chance of success. At the time of the experiment, with two feeds in place and with a radiometer which switched between them so that the beam of the telescope was alternately on a target star and on the sky beside the star, what was seen when the beam was in one position was compared with what was seen when it was the other. This was a way of discriminating against terrestrial interference which, of course, could fool an observer into thinking a civilization had been detected. Terrestrial interference will come in both beams, whereas a true signal will only come in one.

Now we have much better telescopes. Figure 8.3 shows the largest, the 1,000-ft telescope at Arecibo, Puerto Rico. It is now only 10 yr after Project Ozma, but we have made a giant step in those years. The collecting area of this telescope is 150 times that of the telescope used in Project Ozma. The total collecting area is 22 acres, which is about equal to the combined collecting area of all the radio and optical telescopes ever built. Not only is it 150 times greater, but our receivers are now some 10 times better, so that in fact today, only 10 yr later, a search some 1,500 times more sensitive than Project Ozma could be mounted. Nevertheless, given such large telescopes, the search will be very lengthy indeed because one must look at many stars and, as we shall see, perhaps at many more frequencies than the optimum frequency, and the result is that the predicted time scales for a true, effective search for extraterrestrial life are of the order of tens of years.

At first glance Fig. 8.4 appears to be a signal from another extraterrestrial intelligent civilization. This is a recording made with the Arecibo telescope. We see a pulsed signal coming indeed from deep in the cosmos far outside the solar system, from a distance of about 1,000 light-years—the

Figure 8.3   The 1,000-ft telescope at Arecibo, Puerto Rico.

likely distance to the nearest civilization. It looks like an intelligent signal, a regular pulsing, in fact a pulsing so regular that it mimics the accuracy of atomic clocks. It is the signal of a pulsar, nature's magic super clock. This signal is exactly what Arecibo would look like to itself if the two instruments were separated by about 1,000 light-years. Indeed, if there were another Arecibo within 1000 light-years of us radiating at earth a pulsed signal, the record would be this conspicuous, this easy to detect.

Figure 8.5 shows the fine details in pulsar pulses. It should give you a feeling for the amount of information that could be recovered from these pulses. Here we see not the whole pulse train, but one pulse after another recorded in detail from the same pulsar as shown in Fig. 8.4. Now you can see that within the pulses there is detail which can be detected with existing telescopes. Therefore, information coded on interstellar signals could be recorded with existing telescopes.

In addition, there are many special problems which must also be considered in developing search strategy. One, of course, is that presented by the civilization which is not intentionally trying to signal us. Can we do something about it? We will return to that question, but first let us consider the possibility suggested by Von Horner several years ago that interstellar communication may consist of a two-stage process. The first

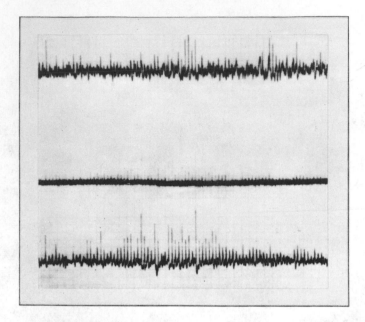

Figure 8.4 Pulsed signal recording made with the telescope at Arecibo, Puerto Rico.

stage, the so-called beacon signal, consists of some method which is designed to call our attention to the existence of a civilization and which will indicate where to look. This, of course, greatly simplifies our search. A second phase is a communication which contains useful information. We should, therefore, keep in mind that there may exist signals which carry no information other than the fact that a civilization exists and that we should pay attention to that part of the sky. For example, a beacon signal could be constructed by inserting an artificial spectral line in the light of a star. For a material with a strong atomic transition probability placed in orbit around, say, the sun, it would only require about 100 tons of material, which existing rockets could lift, to insert a detectable artificial spectral line in the light of a star. By detectable is meant a 1 percent intensity line.

Another procedure might be followed in the case where a civilization does engage in space travel (although we have argued against that) and comes to a planet like earth before the civilizations here had perhaps arisen or at least become intelligent. They could leave behind some artifact which would some day be discovered by the intelligent civilization eventually emerging, and be used either for interstellar communication or to gain information immediately. That particular concept was used by Stanley Kubrick in his movie *2001* which is based on the idea that one of the ways of signaling is to leave something behind on earth, but

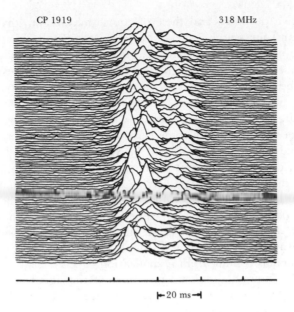

CP 1919   318 MHz

⊢ 20 ms ⊣

Figure 8.5   Fine details in pulsar pulses.

to conceal it so that it would not be discovered until the civilization had evolved to such an extent that it would appreciate the significance of this thing when found. If such things exist anywhere in the universe, one expects them in fact not to be conspicuous but to require sophisticated technology for their detection. One possibility is to hide such artifacts in the interior of limestone caves, regions which will last for perhaps millions of years undestroyed by the environment. One could imagine, for example, hiding such a thing inside the wall of a cave with a radioactive element present so that this thing could be detected only by its radioactivity and not by the inspection of some primitive man. Kubrick's idea went a step further and put the object on the moon, so that the civilization had to be at least sufficiently sophisticated to have space travel before it could find the artifact.

The more general problem of civilizations which are not transmitting special signals for our consumption is, in fact, a very realistic one because, after all, we do not send signals, and the possibility that everybody is listening, and nobody is sending must be considered. In fact, we will not send signals until we have received them, because we do not know in which direction to send. It would be far too expensive to construct enough radio telescopes to bathe the sky in detectable signals, at least in our civilization as we now know it. If a civilization is not transmitting signals, we have to somehow tune in on the communications they use for their own purposes, and these of course may be on any frequency, which means our search is very much more difficult. First, of course, we cannot just look at the economical frequencies or the magic frequencies. We must look at all frequencies. In addition, we may expect all the signals to be weaker because they may not be beamed in our direction. Television, for instance, is beamed all over the sky; whereas the signals we were discussing earlier are focused with a tremendously strong gain of the order of a million in some cases in our direction—or so we have hypothesized. Thus the signals will be very much weaker, and this of course cuts down the range at which we can detect other civilizations. The problem is to devise some way to overcome this difficulty, at least partially. What we would like to do in fact is detect not the individual strong signals, but rather the total ensemble of signals that the civilization is using for its own purposes. With this as the starting point, we can construct ways to detect not individual signals, but the existence of a system of signals coming from a civilization.

Figure 8.6 shows a simulated recording of the signal received in the normal way from a civilization, where we imagine that we have pointed our radio telescope at the distant civilization and have slowly tuned our radio receiver across the band just as you do when you are searching for a radio station. This is a simulation, of course, prepared on a computer. In this case, 400 frequencies are present; 400 distinctive bands have been sequentially sampled. The curve we see simply marks the intensity of the output of our radio receiver, and in this case all the peaks and valleys are almost entirely the result of the noise in the receiver, as we discussed earlier. This is, in fact, not a single such recording but the mean of two recordings in which the receiver has been scanned twice across in frequency. The two recordings have been added together, with the frequencies synchronized so that the data for one frequency is added from one record to the data from another record at the same frequency. In this recording we have taken an ordinary gaussian noise distribution and have

Figure 8.6 A simulated recording of the signal received from a civilization.

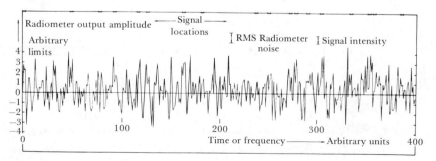

added 18 signals whose locations can be seen. The strength of those signals is shown and the mean-square radiometer noise is given also, and as you see, each of these signals is less strong than the noise in the system. This, of course, is the reason that when you look at this recording you have no impression that you have detected intelligent civilizations. In fact, in some cases the signals do lead to peaks, but in others, they add to the noise in such a way that they do not produce a conspicuous peak.

This discussion has presented a standard approach that is immediately apparent for conducting all these searches and for proceeding to search for the single, strong interstellar signal. However, there is a procedure we can follow with two such records in order to detect the ensemble of signals even though no individual signal is itself detectable. The two recordings can be cross-correlated mathematically. The data from both records at the same frequency are multiplied, the products added, then the two recordings are moved with respect to one another by one unit, the procedure is repeated, the sum is taken, and so forth. Therefore, where there are signals present on both records, as the recordings are multiplied together, the product becomes larger. Thus, when the two recordings are perfectly aligned, the result of such a simple multiplication and addition process is to produce a larger sum. When this is done and the sum of the products is plotted versus the shift of one record with respect to the other, a particularly large maximum is expected when the shift is zero. The mathematics can be worked out, and it can be shown that indeed such a procedure permits the detection of the ensemble of signals even though no individual signal is detectable. If we put N as the number of bands sampled, m as the number of bands with signals, ε as the RMS signal strength, and σ as the RMS noise, then the cross-correlation will have a distinctive observable peak if the following inequality is fulfilled.

$$m/\sqrt{N} > \sigma^2/\epsilon^2$$

The cross-correlation curve for this example is fixed so that the inequality is achieved by only a factor of two. Figure 8.7 shows the cross-correlation curve where the marginally detectable (as it should be since the inequality is just barely fulfilled) peak at zero can be seen. Thus by using this cross-correlation process those records that looked merely like noise indeed can produce evidence that there is an ensemble of signals present. There is, then, at least one procedure, the cross-correlation technique, by which we can detect civilizations even though none of the signals are individually detectable.

To realize the importance of this procedure, consider that if we sample the entire radio window of 10^{10} cycles/sec, with 100-cycles/sec bands, we sample 10^8 frequencies. That is a lot of frequencies, but the task is easier to do than it was before the advent of wide-band receivers, auto-correlators, and similar devices. Let us say that half of the spectrum contains

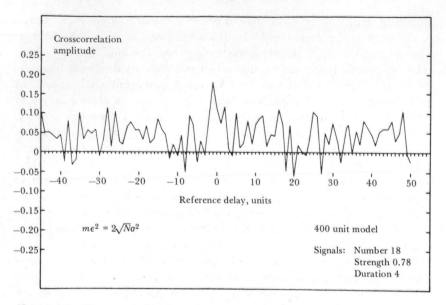

Figure 8.7   Cross-correlation curve.

signals; that is not at all unreasonable because that is true of the earth already. We find that we can detect the existence of a civilization if the signal-to-noise ratio is 1 percent, in other words, if the typical signal strength was only 1/100 as strong as the noise of the receiver. Thus, although every signal present was 100 times weaker than detectable, we could still detect the existence of that civilization. That means that by using this technique the range to which we can reach is increased by a factor of 10 for that particular example; for the number of stars within range, by a factor of a thousand. Therefore this is in fact a very promising technique. To increase the probability of detecting another civilization this technique must be used because it is a general technique which will work no matter what other civilizations are doing. The earlier technique of using certain special signals depends upon the civilization being cooperative. Of course, we are also detectable to other civilizations. The radio signals of mankind, strong enough to detect, are now approximately 40 or 50 light-years out in space and moving outward. We have already become detectable to a large number of stars.

One last and very important problem to consider is that of language if we should find a specific interstellar message. Using the technique just described, of course, we find the messages these civilizations use for their own purposes. These might include television signals which would give us information right away with no need for an interchange. In fact we get the answer immediately. But suppose that we get a purposeful interstellar communication designed by a civilization for the consumption of other civilizations. Could we make anything of it? There has been concern that because there has been the historical precedent of Egyptian hieroglyphics, in which we had countless messages which could not be deciphered until we found the Rosetta Stone, the same situation would occur with a communication in space. We would have the frustrating situation of receiving a message and not being able to make out what it means. That would probably be the greatest frustration in the history of mankind.

There are solutions to this problem, however. For one thing, we do have things in common with the other civilizations. For example, the laws of physics and the arrangement of the universe. These do provide us with a common basis from which we can construct a language. It is possible to construct a language from scratch using the concepts of mathematics, for example. A message can be sent which says "2, some symbol, 2," and another symbol, "4," and we would soon deduce that one symbol equals + and the other means =, and so on. In fact there is a book in existence called *Lingua Cosmica* by Hans Freudenthal which starts from such a premise and develops a very sophisticated interstellar language. By building in such a way, he is able to express such things as emotions over the interstellar communication link by so building logically from the simplest concepts. The problem with *Lingua Cosmica* is that if you do not start with the first chapter, you are completely lost. That is also the problem with such languages. If anywhere along the way the message is missed for a day because there was a brown-out and your telescope went out of operation for a day, you cannot understand any of the subsequent messages. Therefore, we look for other methods— methods which might be called anticryptography. They are codes that are designed to be broken rather than to conceal information.

We shall examine one example of a code which can be broken without any prior communication between the parties concerned. Not only can it be broken easily, but it contains a great deal of information right from the start, which eliminates the thousand-year delay time in transmitting information. A remarkable thing about this message is that despite the fact that this is the first message between two civilizations, designed to be easy to decode, the information content exceeds Shannon's limit. There is more information present than information theory says can be present. The interstellar message is shown in Fig. 8.8. All 551 characters are written here sequentially as zeros and ones which simply means that the message came in simple form, something like a binary code. It might have consisted of pulses with spaces between the pulses, or dots and dashes, or phase-shift modulation, or two tones. There are only two characters used in the message. The problem now is to decode it. The solution should be unique. If we could get more than one plausible decryption of such a message, it is not a very good coding system.

```
1 1 1 1 0 0 0 0 1 0 1 0 0 1 0 0 0 0 1 1 0 0 1 0 0 0 0 0 0 0 0 1 0 0 0 0 0
1 0 1 0 0 1 0 0 0 0 0 1 1 0 0 1 0 1 1 0 0 1 1 1 1 0 0 0 0 0 1 1 0 0 0 0
1 1 0 1 0 0 0 0 0 0 0 0 1 0 0 0 0 0 1 0 0 0 0 1 0 0 0 0 1 0 0 0 1 0 1 0
1 0 0 0 0 1 0 0 0 0 0 0 0 0 0 0 0 0 0 0 0 0 0 0 0 1 0 0 0 1 0 0 0 0 0 0
0 0 0 0 1 0 1 1 0 0 0 0 0 0 0 0 0 0 0 0 0 0 0 0 0 0 1 0 0 0 1 1 1 0 1
1 0 1 0 1 1 0 1 0 1 0 0 0 0 0 0 0 0 0 0 0 0 0 0 0 0 0 0 1 0 0 1 0 0 0
0 1 1 1 0 1 0 1 0 1 0 1 0 0 0 0 0 0 0 0 1 0 1 0 1 0 1 0 1 0 0 0 0 0 0
0 0 0 1 1 1 0 1 0 1 0 1 0 1 0 1 1 1 0 1 0 1 1 0 0 0 0 0 0 0 1 0 0 0 0 0 0
0 0 0 0 0 0 0 0 0 1 0 0 0 0 0 0 0 0 0 0 0 0 1 0 0 0 1 0 0 1 1 1 1 1
1 0 0 0 0 0 1 1 1 0 1 0 0 0 0 0 1 0 1 1 0 0 0 0 0 1 1 1 0 0 0 0 0 0 0 1
0 0 0 0 0 0 0 0 0 1 0 0 0 0 0 0 0 0 0 1 0 0 0 0 0 0 0 0 1 1 1 1 1 0 0 0 0 0
0 1 0 1 1 0 0 0 1 0 1 1 1 0 1 0 0 0 0 0 0 0 1 1 0 0 1 0 1 1 1 1 1 0 1 0
1 1 1 1 1 0 0 0 1 0 0 1 1 1 1 1 0 0 1 0 0 0 0 0 0 0 0 0 0 0 1 1 1 1 1 0
0 0 0 0 0 1 0 1 1 0 0 0 1 1 1 1 1 1 1 0 0 0 0 0 1 0 0 0 0 0 1 1 0 0 0 0
0 1 1 0 0 0 0 1 0 0 0 0 1 1 0 0 0 0 0 0 0 1 1 0 0 0 1 0 1 0 0 1 0 0 0 1
1 1 1 0 0 1 0 1 1 1 1
```

Figure 8.8   A hypothetical interstellar message in binary code.

At first glance the message looks like a telegram, a long sentence. This is where our thinking is already limited; we must broaden our minds. Messages can be one-dimensional, two-dimensional, or three-dimensional; in fact the messages we see with our eyes are typically three dimensional, indeed, four-dimensional. There is color, as well as three spatial dimensions. Therefore, the first thing we must realize is that we cannot assume that the message is one-dimensional but rather we must seek its dimensionality. One way is by trial and error, and we will examine another way below. In this case there is an easier method. Note the number 551 is the product of 29 and 19, both of which are prime numbers measuring that this equation can be factored into 29 groups of 19 or 19 of 29. However, it cannot be factored into a three-dimensional array. Thus, we have been conveniently told that the message is two-dimensional. In fact, what we are dealing with here is the equivalent of a television picture. This is not indeed a subtle way of constructing a code; it is the way human beings and probably many other creatures learn how to talk. One of the simplest modes is that of a picture.

Knowing then that we should break it up either into 19 groups of 29 or 29 of 19, then by trial and error we find which gives a result which looks sensible. Without a Rosetta Stone or computer or other aid we can decode the message in 5 min. The right answer is 29 groups of 19. Figure 8.9 shows the proper layout of the information given in Fig. 8.8. Note that all the 1's have been replaced by light areas.

The result is that certain improbable patterns appear which tell you that you have indeed found the right decryption. However, then the message must be understood. Let us examine it and try to explain it as it would be if such a message arrived and were looked at by the world's experts.

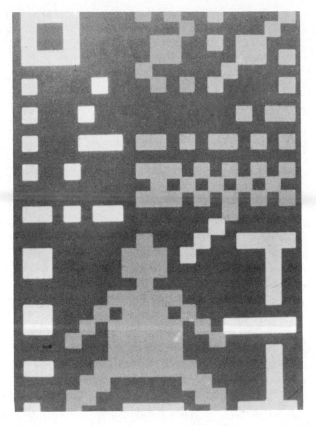

Figure 8.9 The proper layout of the information given in Fig. 8.8.

Indeed if such a picture is given to experts, they identify those parts of the message associated with their specialty but normally fail completely on the part that they are not expert in. This shows that if such a message were received, it would have to be studied by a large number of people before it was fully understood.

The one thing you see here which we probably all understand is the character in the bottom center. This is clearly a drawing of a creature. He is telling us what he looks like. In this case, we see to our amazement that he is a primate much like ourselves with two arms and two legs. He is sort of squatty and fat, which may mean that the gravity on his planet is a little stronger than on the earth, and he appears to have a pointed head. We actually now know in some detail the sort of creature who is communicating with us.

The next easiest thing to recognize is the object in the upper left-hand corner which the astronomers recognize as a diagram of a planetary system. It shows a sun, four minor planets, an intermediate planet, two major planets, an intermediate planet, and a minor planet. The system is very much like our own and in fact is much like what all of astrophysical theory says other planetary systems should be. So we now know the basic arrangement of his solar system. In the upper right-hand corner we see characters which chemists would recognize: a diagram of the carbon atom with its nucleus, two inner electrons, and four outer electrons; and the oxygen atom, with two inner electrons and six outer electrons. They are telling us that they are carbon-based life like ourselves and that they use oxygen. Thus their basic biochemistry is the same as ours.

The hardest group to interpret, surprisingly, is the group which is the key to the rest of the message. That group is nothing more than the numbers 1, 2, 3, 4, and 5 written in binary code but with parity bits added as we add them in computers. They are added in computers to make sure the computer is working correctly. It is anthropomorphic to conclude that the parity bits were added to ensure us that the message-sending system is working correctly. Rather, there must be some other reason which we must guess. What we guess is that the parity bits were added so that the number of characters is always odd, which is what the parity bits do. Thus the numbers will always have an odd number of characters and anything with an odd number of characters is a number. The message is using binary code—the simplest number system.

This deduction tells us that the character at the bottom is not a number. It is probably a word, and we see that it is adjacent to the drawing of the creature. We can not be sure until we see subsequent messages, but what he is probably giving us is the name for himself so that in future messages he will not have to use much message to describe the first person singular. Now that we know which are numbers and which are words, we can see that there is a diagonal line associating some numbers with this creature. The numbers are written alongside certain planets. These are the numbers 5, 2,000 and approximately 4 billion. It is a good guess, though of course we are not certain, that he is telling us his population, so we deduce that there are 4 billion creatures like himself on his home planet; a colony of 2,000 on planet 3, and a scientific expedition group of 5 on planet 2.

Now we come to the last group which extends from his feet to his head and includes a group which is the number 31. He is telling us his size. He is 31 somethings tall. At this point there is a common unit of length which is the wavelength on which the message arrived. If it arrived on the optimum frequency of about 10 cm, he is 10-ft-tall.

Thus, not only can we decode it, we can learn a great deal of useful information from this message. Note that it took only a short time to explain the information in this message, yet according to Shannon's theorem, 551 bits is the equivalent of 25 English words. There are much more than 25 English words in this message, so it does violate, in a good sense, Shannon's criterion for channel capacity. It does that because the message is in fact giving us a coded symbol which we compared against a catalog of possible interpretations. The message does not describe in detail the carbon atom. We compare it with our catalog of possible atoms and pick out the right one. Thus we reach the answer carbon atom without using the number of bits that would be required to spell out carbon atom in English. That is why, of course, the same thing appears elsewhere in the message, and it is why the message can be deciphered.

The business of decoding messages has now been made more mathematical by the Russians who in fact have a group working in the Institute of Linguistics in Moscow to develop mathematical methods for the decryption of such messages.

The Russians have invented a criterion to be used by computers to find the dimensionality of the correct decryption of these messages. In doing this they use a very simple concept which is that when you have the right arrangement of the characters in lines and columns and rows or whatever, you will see a maximum number of straight lines. They may be horizontal, vertical, or diagonal. The sum of the incidences of straight lines, which is called U_l, is in a way a measure of the randomness of the message. If we have the right decryption, there will be more straight lines, and this number will become larger. This is a criterion one can use. One can try all sorts of possible arrangements on the computer with each one computing U_l, and the one with the highest number is the right answer, even though the picture you get may be a very difficult one to understand. In this way you can determine that you have the right answer. Thus, not only are there means to create anticryptography, but indeed the first steps have been made to automated, computerized systems for decoding messages.

We can now conclude that electromagnetic radiation is indeed the most probable system of interstellar communication simply because it is most economical. But we cannot be led to a precise frequency for search simply because we are not sure that civilizations will overtly try to signal us. Nevertheless, a potentially successful search is possible because our present equipment is capable of detecting reasonable signals over the great interstellar distances of 1,000 light-years. The search will be a very lengthy one and will take a large number of resources and great determination and hard work. But once we have the message, as you have just seen, the rest is easy.

A Concrete Plan

The Ames Lecture Series upon which these chapters are based was held in the summer of 1970. At that time, Dr. B. M. Oliver gave a lecture upon the subject of the technology of interstellar communication and discussed what might be achieved in this field. However, in the summer of 1971 at the Ames Research Center, a summer study was held on technological problems in interstellar communication, which was called Project Cyclops. The following chapter by Dr. Oliver represents a drastic updating of material, which incorporates the results of the Cyclops summer study. The detection system which he outlines, which is possible to build with the present technology, but which is admittedly expensive, is truly breathtaking in its capabilities. If human civilization is to develop upon a rational basis, then sooner or later a detection system of the type described in the next chapter must be built in order to search for interstellar communications or inadvertent signals.

This chapter is more technical than the others in this book, and not all readers will be able to follow it. Those who can will find it to contain a masterful summary of the technical arguments that have led to the detailed proposals in the Cyclops study. A brief summary of the essential points is given here.

The same expenditure of energy which would be needed for a manned interstellar trip would suffice to operate a 1,000-megawatt microwave beacon for 30 million years. This could signal our existence to a million possible nearby stars. The electromagnetic spectrum is very wide, but it is much more efficient to use radio microwaves than optical or infrared lasers. The chosen microwave frequency should avoid atmospheric absorption and the frequencies of interstellar molecules, and should be chosen to minimize interference from cosmic radio noise. With directive antennas a 1-megawatt transmitter could then send intelligent signals over a distance greater than 1,000 light-years.

However, to establish a communication link with an extraterrestrial civilization requires that the civilization be attempting to attract attention by means of a radio beacon, radiating in all directions, and we must establish a similar beacon. Optimization then requires the expenditure of about equal amounts of money on the beacon and on the receiving system. However, it makes sense to build the receiver first to see if we can detect any existing beacons.

Intelligent signals from other stellar systems are likely to be contained in a very narrow bandwidth in the microwave spectrum, but they cannot be made too narrow because of frequency shifts produced by relative motions of the earth and the sending planet (doppler shifts). Recent technical advances now allow the use of an optical spectrum analyzer technique to search simultaneously for narrow-band-width signals over a large part of the microwave spectrum. For various technical reasons the optimum part of the microwave spectrum that should first be subjected to this technique lies between 1,420 and 1,662 MHz.

The Cyclops proposal is for the construction of an array of antennas covering a circular area having a diameter of about 10 km, but with a filled area equivalent to about 10 percent of this. Antenna diameters would be about 100 m. The total cost of such a system would be about 6 billion dollars, but this sum could be spent gradually, with the first part of the array being used to monitor the nearby stars. Frank Drake listened to Tau Ceti and Epsilon Eridani, 10 to 11 light-years from us, in Project Ozma. The Cyclops system would have an equal signal detection efficiency at 2,000 times the distance of Project Ozma, and would be able to detect much weaker signals from closer stars. Beacons of reasonable power could be detected up to 1,000 light-years, a volume of space containing more than 1 million stars fairly similar to the sun.

9

Technical Considerations in Interstellar Communication

Bernard M. Oliver   *Hewlett-Packard Company*

Once our society becomes convinced of the existence of intelligent life elsewhere in our galaxy, we will embark on the greatest voyage of discovery in all our history. Curiosity and the appreciation of the potential cultural and scientific benefits of contact with other advanced species will drive us to make the attempt. Since the popular support for such an effort may be imminent it is timely to consider how we might best do it.

One thing appears certain at this time. We will *not* search the galaxy in spaceships, looking for life; the distances are just too great. Our manned space flights to the moon have taken us 1¼ light-seconds into space; the nearest star is about 4 light-years from us. To find intelligent life, we could have to hunt for favorably situated planets around thousands, and possibly hundreds of thousands, of stars ranging from ten to several hundred light-years from the sun in all directions.

Even a single space flight involves staggering costs in human time or in energy. With foreseeable technology, we could launch a spaceship that would escape the solar system at 18 miles/sec. This is one ten-thousandth the velocity of light, so the journey to Alpha Centauri, the nearest star system, would take 40,000 yr. Clearly we need higher speed vehicles before any astronauts volunteer. Let us discard all limitations of present technology, but not the limitations imposed by known physical law, and assume the best possible rocket spaceship. This ship annihilates matter with antimatter and ejects pure radiation as its exhaust. For such a ship to reach seven-tenths ($\sqrt{0.5}$) the speed of light, the mass ratio must be $1 + \sqrt{2}$. To make the four accelerations needed for a round trip, the mass ratio must be about 34. At seven-tenths the speed of light, the ship will travel one light-year in 1 yr of ship time, so the crew would need provisions for about 10 yr to fly to Alpha Centauri, examine the system, and return. Perhaps a 1,000-ton vehicle would suffice. If so, the takeoff weight would be 34,000 tons, of which 33,000 tons would be annihilated en route. The energy release would be 3×10^{24} joules, or about 10^{18} kilowatt-hours. *Enough energy would be squandered on one mission to keep a 1,000-megawatt microwave beacon operating for 30 million years.* Such a beacon, as we shall see, could signal our existence to *a million* likely stars.

Even disregarding the energy cost, other problems arise. The power generated, even in a leisurely takeoff from an orbiting station would be about 10^{18} watts. If only one-millionth of this energy (gamma rays, by the way) reaches the ship, the heat flux alone would be 1 million megawatts. To radiate this much heat at room temperature, we would need about 1,000 square miles of radiating surface. And, of course, we would have to do something about interstellar dust, each grain of which becomes a miniature atomic bomb when intercepted at nearly the velocity of light. Unforeseeable breakthroughs must occur (and may never occur) before man can physically travel to the stars.

The facts are that to send sizeable masses over interstellar distance in reasonable times is enormously expensive, even with any technological advance we can foresee. While this may be a disappointing conclusion for the science fiction fan, it may also lend comfort to those who fear interstellar attack as a consequence of contact with other races.

Interstellar Communication

Philip Morrison has estimated that all we know of ancient Greece could be contained in about 10^{10} bits of information, a unit he suggests be named the Hellas. For two advanced civilizations to describe themselves and their knowledge to each other might involve the exchange of something on the order of 100 Hellades of information. This is vastly less expensive than exchanging tons of metal.

To receive information we must, at the very least, be able to detect the presence or absence of a signal. To detect reliably a pulse (representing the presence of the signal) the expectation, or average, particle count per pulse must be sufficiently high so that the probability is low of receiving a number of particles that does not significantly exceed the natural background. Thus to conserve power

1. The energy per particle should be as low as possible.
2. The particle velocity should be as high as possible.
3. The particles should be easy to generate, launch and detect.
4. The particles should not be deflected by fields in space.
5. Absorption by interstellar matter should be negligible.

Except for photons, uncharged particles are difficult to accelerate, direct, and detect. Charged particles are deflected by the galactic magnetic field and, except at *very* high energies, do not penetrate atmospheres. The *kinetic* energy of an electron travelling at half the speed of light is already ten billion times the total energy of a 1,420 MHz photon. Photons are as fast as any known particle, are easily generated and detected, suffer no deflection, and (at microwave frequencies) have negligible absorption, and are the least energetic. There seems little doubt that electromagnetic waves are the only really suitable means for interstellar communication, and that communication is our only hope of initial contact.

Some Fundamental Relations

If a transmitter were to radiate its power P_t isotropically then, at a distance R, the power would be spread over a sphere of area $4\pi R^2$. A receiving antenna of collecting area A_r at the distance R would therefore collect a power

$$P_r = \frac{P_t A_r}{4\pi R^2}. \tag{9.1}$$

If the power gain of the transmitting antenna in the given direction is g_t, the received power will be increased by this factor, that is,

$$P_r = \frac{P_t g_t}{4\pi R^2} \; A_r = \frac{P_{\text{eff}} A_r}{4\pi R^2}, \tag{9.2}$$

where $P_{\text{eff}} = P_t g_t$ is the "effective radiated power."

The gain of a receiving antenna is obviously proportional to its area and, by reciprocity, the same is true for a transmitting antenna. It can be shown in many ways, including thermodynamically[1] that

$$g = \frac{4\pi A}{\lambda^2}, \tag{9.3}$$

[1] B. M. Oliver, "Thermal and Quantum Noise," *Proc. IEEE*, **53**, 5, 442.

where λ is the wavelength of the radiation used. Thus Eq. (9.2) may be written in several equivalent forms, for example,

$$P_r = P_t \frac{A_t A_r}{\lambda^2 R^2} = P_t \left(\frac{\lambda}{4\pi R}\right)^2 g_t g_r. \tag{9.4}$$

For given antenna *areas*, short wavelengths are best; for given antenna *gains*, long wavelengths are best.

Antenna gain and directivity are inseparable. A transmitting antenna achieves its gain by concentrating its radiation in the desired direction at the expense of other directions. The same antenna used as a receiver has high gain (and collecting area) in one direction at the expense of gain (and collecting area) in other directions.

For an antenna having a circular collecting area of diameter d, we may take $A = \pi d^2/4$ in Eq. (9.3) and find that

$$g_o = \left(\frac{\pi d}{\lambda}\right)^2, \tag{9.5}$$

where we have written g_o for g to signify the on-axis gain. The gain in other directions is then

$$g = g_o \left[\frac{2 J_1(\pi d\theta/\lambda)}{\pi d\theta/\lambda}\right]^2 = g_o \left[\frac{2 J_1(\theta\sqrt{g_o})}{\theta\sqrt{g_o}}\right]^2, \tag{9.6}$$

where θ is the angle off axis and J_1 is the first order Bessel function. The expression in brackets is unity for $\theta = 0$ and drops to the value $1/2$ when

$$\theta = \theta_{1/2} = \frac{1.616}{\sqrt{g_o}} = (0.5145\ldots)\frac{\lambda}{d} \tag{9.7}$$

Thus the bandwidth between half-power points is

$$2\theta_{1/2} = (1.029\ldots)\frac{\lambda}{d} \approx \frac{\lambda}{d}. \tag{9.8}$$

The number of directions in which we must point an antenna to search the entire sky is proportional to its gain. An isotropic antenna has a gain of one and must be "pointed" in only one direction. An antenna radiating uniformly into a hemisphere has a gain of two and must be pointed in two directions, and so on. Because the gain patterns of actual antennas are not constant over a sector, the number of pointing directions needed depends on the loss of gain that we are willing to tolerate at the edge of the field. If we somewhat arbitrarily set this loss at 1 db, then it can be shown from Eqs. (9.5) and (9.6) that the number of directions, N, is

$$N \approx 4g. \tag{9.9}$$

This relation has an important bearing on our search strategy as we shall see.

The Microwave Window

Any receiver of electromagnetic radiation has an irreducible amount of noise, which, referred to the input, has the spectral power density

$$\psi(\nu) = \frac{h\nu}{e^{h\nu/kT_S} - 1} + h\nu, \tag{9.10}$$

where ν is the frequency, h is Planck's constant, k is Boltzmann's constant, and T_s is the absolute temperature of the source. The first term in Eq. (9.10) represents black-body radiation (in a single propagation mode). If the receiver is coherent, that is, if it amplifies the received signal and preserves its phase, the second term is caused by spontaneous emission in the amplifier. If the receiver is incoherent, that is, if it detects only the energy of the signal as a photocell does, then the second term represents statistical fluctuations in the photon count and can be considered to be "photon shot noise."

At high frequencies such that $h\nu/kT_s >> 1$, (that is, well beyond the quantum cutoff) the first term disappears and $\psi(\nu) \to h\nu$. The receiver noise power is then proportional to frequency. This simply reflects the increasing energy per particle, discussed earlier. At low frequencies such that $h\nu/kT_s << 1$,

$$\psi(\nu) \to kT_S + h\nu \approx kT_S \tag{9.11}$$

and the noise in an ideal receiver is proportional to the source temperature. Since this is the usual situation in radio and microwave systems, it has become customary to describe receiver noise in terms of an equivalent noise temperature

$$T = \frac{\psi(\nu)}{k}. \tag{9.12}$$

When the receiver is not ideal, its excess noise is accounted for as an equivalent increase in the source temperature. By using cryogenically cooled amplifiers and low-loss antennas and waveguides, the receiver noise-temperature increment can be as low as $16°$K.

When a microwave antenna is pointed at the sky, it sees several sources of noise, whose noise temperature vs. frequency are shown in Fig. 9.1. The first of these is the isotropic background radiation, believed to be the adiabatically expanded radiation of the original "big bang." Thus the sky in all directions has a minimum noise temperature of about $3°$K. The curve shown includes the spontaneous emission or quantum shot noise of an ideal receiver operating from a $3°$K source; that is, we set

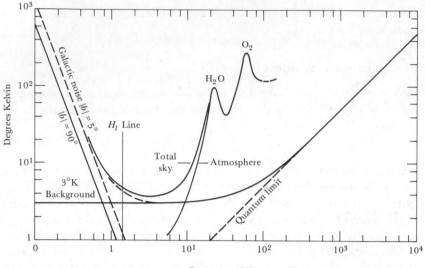

Figure 9.1   Noise temperature versus frequency for several sources of noise.

$T_s = 3°K$ in Eq. (9.10) and find the temperature from Eq. (9.12). The second important noise source is synchrotron radiation from electrons whirling around in the galactic magnetic field. This "galactic noise" was the radio astronomer's original signal, but for interstellar communication it represents an interfering noise. It rises steeply with decreasing frequency below 1 GHz and continues to increase until the ionosphere becomes opaque. Thus there is a broad valley of low noise temperature from about 1 to 100 GHz that is the same for all receivers in space in the solar neighborhood or similar regions of the galactic disk. For a ground-based receiver on any planet with an earth-like atmosphere, the absorption lines of water vapor and oxygen add a third source of noise that becomes significant above 10 GHz. This narrows the valley to a region from 1 to 10 GHz. We shall call this part of the spectrum the "microwave window." For several reasons, given later, the *low end* of this window is best suited for interstellar communication so the possibility that some inhabited planets might not have strong water vapor and oxygen lines is of no concern.

The microwave window also contains noise peaks that arise from spectral lines of elements and radicals in space. The hydrogen line is at 1,420 MHz, and four closely spaced hydroxyl lines cluster around 1,680 MHz, the lowest being at 1,662 MHz. These lines may have great significance for interstellar communication, as we shall discuss later.

The black-body and coronal radiation of normal stars produce an elevation of the system noise temperature that is proportional to antenna area.

In the microwave region for the largest antennas we will consider, this noise-temperature increase is at most a few degrees for the nearest stars. It decreases inversely with the square of distance. Thus, while many normal stars can be detected on microwave frequencies by using large antennas, wide bandwidths, and long integration times, star noise poses no problem for interstellar communication in the microwave window. Star noise is a problem at infrared and optical frequencies.

Range Equations

The range limit for a communication system depends on a number of additional factors such as the type of coding used, the detection method, the acceptable noise level or error rates in the output, and so forth. In a detection or search system the major factors are the integration times and allowable probabilities of false alarms and detection failure. However, a rather natural and useful reference range limit is that for which the received signal power and noise power are equal. Let us therefore derive some range expressions for this case and discuss some of the other factors later.

Since the total received noise referred to the input is kTB, where B is the receiver bandwidth, we find from Eq. (9.4) that the reference range limit R_o is

$$R_o = \frac{1}{\lambda}\left[\frac{P_t A_t A_r}{kTB}\right]^{1/2}. \tag{9.13}$$

The antenna cost may be considered proportional to $A_t + A_r$ and, for a constant product $A_t A_r$, is minimized if $A_t = A_r = A$. Thus, if we have cooperation at both ends and can optimize the system, we have

$$R_o = \frac{A}{\lambda}\left(\frac{P_t}{kTB}\right)^{1/2} = \frac{\pi d^2}{4\lambda}\left(\frac{P_t}{kTB}\right)^{1/2}, \tag{9.14}$$

where the second equation applies if the antennas have a circular aperture of diameter d.

One of the most efficient ways to send information is to encode it in binary form, and to reverse the phase of the carrier to distinguish between a "zero" and a "one." This is the so-called symmetrical binary channel, and the modulation is often referred to as phase shift keying (PSK). The matched receiver will have a bandwidth equal to the bit rate so the radiated energy per pulse is $W = P_t/B$. Thus Eq. (9.14) can be written

$$R_o = \frac{A}{\lambda}\left(\frac{W}{kT}\right)^{1/2} = \frac{\pi d^2}{4\lambda}\left(\frac{W}{kT}\right)^{1/2} \tag{9.15}$$

Synchronous detection is required to recover the signal, and it produces a pulse of one polarity for a zero and of the opposite polarity for a one. A simple polarity detector then decides which symbol has been received. Because of the noise the recovered symbol amplitudes will have a gaussian distribution about their proper (mean) values. A bit will be misread, producing an error, whenever the noise reverses the polarity of the recovered bit. Synchronous detection removes all noise that is in phase quadrature with the received carrier. When the input signal-to-noise ratio is unity (0 db), the probability of a false bit is 0.07865. This is therefore the bit-error rate at the reference range limit. The error rate drops with increasing signal-to-noise ratio (SNR) and can be further reduced by error-correcting codes. Table 9.1 shows the bit-error rate as a function of SNR with and without the use of a good error-correcting code.

Table 9.1

SNR, db	Error rate	
	Uncoded	*Coded*
3	2.3×10^{-2}	1×10^{-3}
4	1.25×10^{-2}	3.2×10^{-5}
5	6×10^{-3}	7×10^{-7}
6	2.4×10^{-3}	1×10^{-8}

We see that at one-half the reference range limit (SNR = 6db) virtually error-free transmission can be realized.

Before proceeding further let us see if we can, in fact, signal over interstellar distances with reasonable powers and antenna sizes. Calculations and experience show that fully steerable antennas 100 m in diameter can be built to operate at $\lambda = 3$ cm. Let us assume a bandwidth of 1 Hz and a system noise temperature of 20°K. Then with a transmitter power of 1 megawatt we find

$$R_o = 1.57 \times 10^{19} \text{ m} = 1,666 \text{ light-years.}$$

At one-half this range we could, with error-correcting codes, achieve virtually error-free transmission at 1 bit/sec. Figure 9.2 shows the range as a function of antenna diameters for various bit rates. We see that with large phased arrays the transmission of high-quality speech and music and even of television is possible over hundreds of light-years. Before all this is possible, however, large antennas must be present at *both* ends of the link precisely pointed in the right directions. In the search phase we cannot count on this.

Lasers vs. Microwaves

The low noise level in the microwave window argues for the use of microwaves for interstellar communication. However, the ease with which

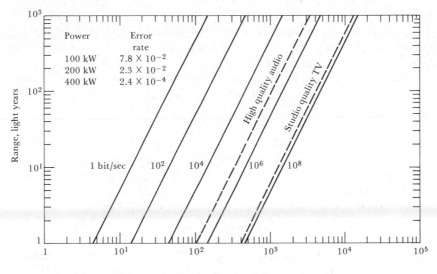

Figure 9.2 A graph showing the range as a function of antenna diameters for various bit rates.

coherent light can be focused into light beams has led many people to propose the use of lasers. To help resolve this question the Cyclops study included a careful comparison of four laser systems and two microwave systems. Without going into all the details we will summarize this comparison here.

Two of the laser systems called Optical A and B operate at $\lambda = 1.06$ μ, the wavelength of the Neodymium glass laser. The other two, called Infrared A and B, operate at $\lambda = 10.6$ μ, the wavelength of the CO_2 laser. Optical A uses short high-power pulses characteristic of the Neodymium glass laser. All the other systems including the microwave systems A and B assume a CW power capability of 10^5 watts. This is beyond the present state of the art for lasers having the spectral purities needed and is well within the state of the art for the microwave systems, especially for microwave B, where a transmitter used with each antenna in the phased array would permit a total power of several hundred or even thousands of megawatts.

The beamwidth of all the laser systems is 1 sec of arc. This is about the narrowest beam that can be used in practice because of atmospheric turbulence and pointing errors. All the laser-transmitting antennas (telescopes) have the diameter required to give this beamwidth. All the receiving antennas except Infrared B have a collecting area equivalent to a 100-m diameter aperture. This is accomplished by adding the output of smaller units after square law or photon detection. Infrared B assumes coherent (synchronous) detection, which limits the collecting area to that of a single element having a 1-sec-of-arc beam.

Microwave system A also uses a 100-m-diameter collecting area, while microwave B uses an array 6.4 km in diameter (3 km equivalent clear aperture) to achieve the 1-sec-of-arc beamwidth. Microwave A thus provides a comparison based on the same collecting area, while microwave B provides a comparison based on the same beamwidth.

The microwave systems and Infrared B assume synchronous detection and a 1-Hz effective bandwidth. It is extremely doubtful that the spectral purity of a few parts in 10^{14} needed for this synchronous detection can ever be realized in a 100-kW laser at $\lambda = 10.6 \ \mu$. Both infrared systems use optical heterodyning followed by intermediate frequency amplification. In Infrared A a 3-kHz IF bandwidth is assumed, which requires the difficult but perhaps not impossible frequency stability of a part in 10^{10}. The IF signals are square-law detected and added.

Optical systems A and B use silicon avalanche photo detectors. In A the predetection bandwidth is the reciprocal of the pulse duration, while in B it is limited by the highest Q (10^8) Fabry Perot filters considered to be practical. It is worth noting here that doppler drift rates, discussed later, make all the laser system bandwidths, save that of Optical A, appear questionably narrow.

The range limits of the synchronous detection systems are given by Eq. (9.14), while those of the other systems were computed to give the same bit-error rates (0.07865 . . .) using the actual statistics of the square-law-detection process. These range limits are given along with the system parameters in Table 9.2. Microwave system A, in spite of its 64 times broader beam has over an order-of-magnitude greater range than any of the laser systems. Microwave system B, which has the same beamwidth as the laser systems, has 10,000 times the range. With 10 megawatts of total power Microwave B could signal over *intergalactic* distances.

Why do the laser systems show up so poorly? Fundamentally, of course, they have a higher noise temperature because of the higher energy per photon. But there are other problems too. *The sole advantage of lasers, the ability to get sharp beams with small transmitting antenna areas, becomes a disadvantage at the receiver where we are prevented from realizing large coherent collecting areas.* This is the downfall of Infrared system B. In Infrared A, each receiver has its own spontaneous emission noise. When the signals from the n antennas are added incoherently (after square-law detection) the SNR is improved only as $\sqrt{n}$ rather than as n as is true for a phased array. Optical system B does not adequately overcome star noise, while Optical A has too little energy per pulse. To allow for doppler rate, the minimum bandwidth of any system must be proportional to the square root of its operating frequency. This factor is ignored in the comparison between Infrared B and the microwave systems.

If we grant the fact that we can realize, through phased arrays, the same limiting beamwidths at microwave frequencies (and are willing to

Table 9.2

Parameter	Optical		Infrared		Microwave	
	A	B	A	B	A	B
Wavelength	1.06 μ	1.06 μ	10.6 μ	10.6 μ	3 cm	3 cm
Transmitter						
Antenna diameter	22.5 cm	22.5 cm	2.25 m	2.25 m	00 m	3 km*
Number of elements	1	1	1	1	1	900
Element diameter, m	0.225 m	0.225 m	2.25 m	2.25 m	00 m	100 m
Antenna gain	4.4×10^{11}	4.4×10^{11}	4.4×10^{11}	4.4×10^{11}	1.1×10^{8}	9.8×10^{10}
Peak or CW power, watts	10^{12}	10^{5}	10^{5}	10^{5}	10^{5}	10^{5}
Modulation	Pulse	Pulse	Pulse	PSK	PSK	PSK
Pulse duration, sec.	10^{-9}	1	1	1	1	1
Energy per bit, joules	10^{3}	10^{5}	10^{5}	10^{5}	10^{5}	10^{5}
Effective radiated power, watts	4.4×10^{23}	4.4×10^{16}	4.4×10^{16}	4.4×10^{16}	1.1×10^{13}	9.9×10^{15}
Beamwidth, in.	1	1	1	1	4	1
Receiver						
Antenna diameter	100 m	100 m	100 m	2.25 m	00 m	3 km*
Number of elements	400	400	1,975	1	1	900
Element diameter, m	5	5	2.25	2.25	00	100
Atmosphere transmission	0.7	0.7	0.5	0.5	1	1
Overall quantum efficiency	0.4	0.1	0.2	0.2	0.9	0.9
Solar background ratio	1.2×10^{-3}	36	1.7×10^{-3}	6×10^{-7}	—	—
Noise temperature, °K	(13,600)	(13,600)	1,360	1,360	0	20
Effective RF bandwidth	1 GHz	3 MHz	3 kHz	1Hz	1Hz	1Hz
Detection method	Photon	Photon	Square law	Synch.	Synch.	Synch.
System						
Range limit, light-year	26	24	22	41	50	450,000
State-of-the-art?	?	No	?	No	Yes	Yes
All weather?	No	No	No	No	Yes	Yes

*Array spread out to 6.4-km diameter to avoid vignetting

pay for the area) then no laser system can ever hope to compete with microwave systems. This is true not only because of the lower microwave noise temperature, but also because of the narrower realizable bandwidth, and because of the λ^2 in the numerator of the last expression in Eq. (9.4). In fact if lasers had been known for centuries and microwaves had only recently been discovered, microwaves would be hailed as the long-sought answer to interstellar communication.

Problems of the Search Phase

The performance figures for microwave links show that communication is technically feasible over distances of 1,000 light-years or more at rates on the order of 1 bit/sec with antennas that are already in existence. But these figures assume that two highly directive antennas are pointed in the proper directions at the proper times and that a receiver is tuned and phase locked to precisely the proper frequency. The reason we are not already engaged in interstellar communication is that we do not know where to point our antennas nor on what frequency to listen or send. Further, we cannot count on beacons being beamed at us by highly directive transmitting antennas. We may have to detect beacons that are radiated omnidirectionally, or even eavesdrop on signals generated by other races for their own purposes. Such signals will be much weaker and require for their detection enormous collecting areas.

We have only a very rough idea as to how far out into space we must carry the search. The major reason for this uncertainty is that we do not know within orders of magnitude the length of time during which intelligent races radiate powerful signals. If races typically radiate megawatts of power as beacons or for their own purposes for 10^7 yr, the average spacing between radiative races might be less than 100 light-years. If the radiative epoch is 10^4 yr, the average separation is an order of magnitude greater, and we might have to search out to 1,000 light-years. Beyond this range the situation becomes rather bleak. If the radiative epoch is only 1,000 yr, the average separation is about 2,000 light-years so the round trip delay exceeds the longevity of the communicative phase. Nevertheless, we cannot exclude the possibility that very advanced races exist beyond this range who have constructed very powerful beacons and who use them for unknown purposes for very long times.

Since the range over which we must search is so uncertain we can only draw some rather general and tentative conclusions.

1. We should start the search with a modest system capable of detecting beacons out to perhaps 100 light-years.
2. We should expand the system as the search proceeds and continue until success is achieved or until we are able to eavesdrop on unintended

radiation from a range of 100 light-years. The system should then be able to detect beacons of reasonable power at 1,000 light-years range.

3. If technologically feasible, we may want to search for more distant powerful sources and perhaps to scan other galaxies.
4. We will want to radiate as well as receive. We are already radiating leakage signals (our UHF-TV stations) but we will want to consider beacons as well.

The stars most likely to have inhabited planets circling them are main-sequence stars of the stellar types F0 through K5 or K9. Stars larger than F stars do not live long enough for Darwin-Wallace evolution to be effective. Stars smaller than K stars radiate so little energy that a planet, to be sufficiently warm to support life, must be so close to the star that tidal friction will stop its rotation. Within 100 light-years of the sun there are about 2,400 of these likely stars; within 1,000 light-years there are some 1.7 million. The search thus involves a large number of stars. We can afford to spend days searching each of the nearby stars, but if we want to search all the likely stars in the 100- to 1,000-light-year range we can only afford a very short time per star—perhaps 15 to 20 min. This implies that the signals we are looking for must be present continuously or we will very likely miss them. It follows that these signals will very likely be beacons, not leakage signals.

Beacons designed to be "visible" at ranges of 1,000 light-years will probably be radiated omnidirectionally. To do otherwise would require the construction of a million directive antennas, each of which would have to be aimed at the spot in the sky where the proper motion of the target star will carry it in one round trip light time. In other words the transmitted beams would have to be aimed so as to lead the stars. This involves knowing the proper motions of all the million or more target stars and their distances. Suppose that 30-m-diameter antennas were used, at a wavelength of 20 cm. The number of resolvable directions for each antenna would then be about 10^6. With a million antennas operating, the sky will be flooded with beams. This same end result could be achieved with a single omnidirectional radiator without the expense of a million 100-ft dishes. Instead of paying for a tremendous antenna installation, the money should be used for increased transmitter power.

Even with very high-powered beacons the loss of the transmitting-antenna gain means that we will need very large and therefore very directive receiving antennas: phased arrays several kilometers in diameter. The long-range search situation is technologically asymmetrical. We must have large receiving-antenna gain and directivity to

1. Collect enough energy to detect the signal,
2. Exclude local interference,
3. Tell us from what star the signal came.

But the transmitted signal can be, and for long-range beacons, probably will be radiated isotropically. The situation is much like normal vision, where the directivity of our eyes enables us to see interesting distant sources that radiate in all directions, and to exclude nearby stray light.

With an omnidirectional source radiating a power P_t and a receiving antenna of area A_r, the reference range limit is found by replacing A_t by $\lambda^2/4\pi$ in Eq. (9.13) and is:

$$R_o = \left(\frac{P_t A_r}{4\pi k T B}\right)^{1/2}. \tag{9.16}$$

Assuming that kTB is fixed, we determine the range by the product $P_t A_r$. The cost of the beacon is proportional to P_t while the cost of the receiver is largely in the antennas and so is proportional to A_r. The total cost of a beacon *and* a receiver is, therefore

$$C = K_a A_r + K_p P_t, \tag{9.17}$$

where K_a and K_p are constants. If we wish to minimize C keeping $P_t A_r$ constant we find this occurs when

$$K_a A_r = K_p P_t \tag{9.18}$$

or, in other words, when the beacon and receiver costs are equal. There is thus an *economic* symmetry to the search. A rough figure for K_p is $2000/kilowatt. Thus if we are willing to spend $2 billion for receiving antennas, we should plan to radiate about 1,000 megawatts in an omnidirectional beacon. If other races reason this same way and have similar economic factors, then this is the order of magnitude of power we can expect in long-range beacons.

If the receiving antenna is a circular area of diameter d then $A_r = \pi d^2/4$, and Eq. (9.16) may be written

$$R_o = \frac{d}{4}\left(\frac{P}{kTB}\right)^{1/2}, \tag{9.19}$$

where we have dropped the subscript of P_t. Let us now consider how narrow a bandwidth B we can use.

Relative motion of the transmitter and receivers can occur because of (a) radial motion of the other star with respect to the sun, (b) orbital velocity of the earth and the other planet, and (c) rotation of the earth and the other planet. Each of these motions produces doppler shifts of the signal.

Radial motions are essentially constant over long times and therefore produce fixed-frequency offsets of as much as a part in 10^3. The principal effect is to broaden the frequency range over which we must search for a signal known (or suspected) to have been radiated at a particular frequency such as the hydrogen line. Rather than looking at exactly 1,420 MHz, we must scan a band 2.8 MHz wide centered at 1,420 MHz. (For a laser signal at 10.6 μ, the band would be 56 GHz wide, and at 1.06 μ, the width could be 560 GHz).

Orbital motions and planetary rotations produce a nearly sinusoidal frequency modulation of the signal with a peak deviation

$$\Delta\nu = \frac{a\Omega}{c}\nu \qquad (9.20)$$

and a peak rate of change

$$\dot{\nu} = \frac{a\Omega^2}{c}\nu, \qquad (9.21)$$

where ν is the original signal frequency, a is the radius of the orbit or planet, and Ω is the angular velocity of the planet in its orbit or around its axis. For the earth we find that

	$\Delta\nu/\nu$	$\dot{\nu}/\nu$
Orbital motion	10^{-4}	2×10^{-11}/sec
Diurnal rotation	1.5×10^{-6}	1.1×10^{-10}/sec

The frequency offsets are less than we must accommodate because of stellar radial velocities and so do not concern us. The diurnal *rate*, if uncompensated, sets a minimum limit on our bandwidths.

A signal drifting at a rate $\dot{\nu}$ will sweep through a channel B Hz wide in $\tau = B/\dot{\nu}$ sec. Since the response time of the filter is about $1/B$ sec, the limiting bandwidth is that which makes $\tau \approx 1/B$. Thus

$$B_{\text{opt}} \approx \dot{\nu}^{1/2}. \qquad (9.22)$$

Wider bands will admit more noise, narrower bands will not fully respond to the signal.

The earth, because of its relatively large moon, has been slowed appreciably in its rotation. It would not be surprising to discover planets of earth size with 8-hr days and, therefore, values of $\dot{\nu}/\nu$ nine times as great, or 10^{-9}/sec. This would suggest bandwidths on the order of 1 Hz for received frequencies of 1 GHz.

Doppler rates can be compensated in any given transmission or reception direction by drifting the transmitter or receiver local oscillator at the opposite or same rate, respectively. However, an omnidirectional beacon radiated by a single antenna cannot be doppler compensated in all directions at the same time. Omnidirectional beacons are therefore better radiated by a pair of antennas at the poles of the planet or by a ring of east or westward pointing antennas around the equator. Nonrotating beacons in orbit are another possibility that might be used by an advanced race.

Substituting Eq. (9.20) into Eq. (9.17) we find the reference range limit for doppler drifting signals

$$R_o = \frac{d}{4}\left(\frac{P}{kT\dot{\nu}^{1/2}}\right)^{1/2}. \tag{9.23}$$

Figure 9.3 shows R_o as a function of d for various assumed values of $\dot{\nu}$. As a result of doppler compensation we may be able to count on $\dot{\nu}$ being as low as 10^{-2} Hz/sec, particularly at the low end of the microwave window. Even so we would need an antenna diameter on the order of 6 km to receive a 1,000-megawatt omnidirectional beacon at 1,000 light-years. Such an array would also let us detect 10 megawatts at 100 light-years so we can begin to look for leakage signals as well as beacons.

Apparently we are going to need very large coherent collecting areas in the search phase. An array with 3-km clear aperture would have to be spread out over a 10-km diameter area to avoid self-shadowing at low elevation angles. Such an array would have about 10^{11} resolvable directions. The microwave window contains 10^{10} 1-Hz-wide channels. If the signal were beamed at us for a short time, say 1 sec/day, we would have to search each channel for 10^5 sec. To scan the entire sky, blindly, with our big antenna and a single receiver with a 1-Hz bandwidth would require a time

$$T = 10^{10} \times 10^{11} \times 10^5 = 10^{26} \text{ sec} = 3 \times 10^{18} \text{ yr}$$

or about 300 million times the age of the galaxy!

Figure 9.3 R_O as a function of d for various assumed values of $\dot{\nu}$

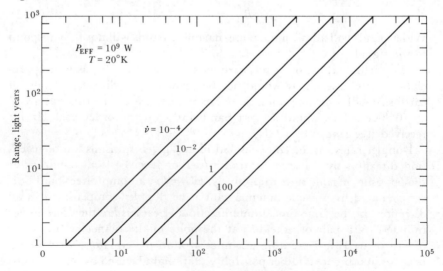

Clear aperture diameter, m

If the array elements are 100 m in diameter, their field of view is 10^4 times as large (in solid angle) as the array beam. This gives 10^7 fields of view in the entire sky or six times as many fields as there are target stars. There would therefore be one-sixth of a target star per field on the average. So we cannot rely on imaging to speed the search. What then can we do to shorten the time required?

First, *we must not search blindly.* We must, by an optical survey, identify the million or more stars of the right spectral range that lie within 1,000 light-years of the sun, and store their coordinates in a large data base. We can then search these in order of increasing range and not waste time on the voids in between. Result: 10^{11} directions → 10^6 directions.

Second, *we ought to search all likely channels in the spectrum simultaneously.* At present we cannot do this for the entire microwave window, so we must decide that there is some narrower natural band and design a system to search this band simultaneously. Result: 10^{10} observations/star → 1 observation/star.

Third, *we must assume that the signal we are looking for is either present continuously, or else comes from a nearby star.* There are only 1,000 stars sufficiently near to permit eavesdropping on their occasional leakage signals. We can afford to spend days on these, and should. But for stars beyond this range we must depend on beacons that are always shining. Result: 10^5 sec → 10^3 sec.

Under these assumptions, the search time drops to 10^9 sec, or about 30 yr. We believe these assumptions are reasonable. We believe that an automated optical search system can be developed that can pinpoint the likely stars both as to spectral type and range. We believe that leakage radiation is much more likely to be detected at the low end of the microwave window and that beacons are likely to lie in a narrow frequency range at the low end of the window. Finally, as a result of the Cyclops study, we believe we can construct receivers that can "comb" 200 MHz of bandwidth into 1-Hz or 0.1-Hz channels; that is, we can construct a receiver with a *billion* or more simultaneous outputs.

Likely Frequencies for Beacons

There are several reasons to prefer the low end of the microwave window for the search phase. These include

1. Smaller doppler drift rates
2. Less stringent frequency stability requirements
3. Broader beamwidths requiring less pointing accuracy
4. Less severe surface tolerances, permitting cheaper antennas
5. Smaller power densities in transmitter tubes, waveguides, and feeds
6. Greater freedom from O_2 and H_2O absorption, and from rain loss (all of which may be more severe on other inhabited planets).

Thus it appears likely that the search system need only cover the window from perhaps 0.5 to 3 GHz. This is still an enormously large bandwidth to be combed with a 1-Hz channel. So the question arises as to whether there is a more sharply defined region of the spectrum where there is a reason, common to all intelligent races, to locate the search.

Several years ago Cocconi and Morrison suggested that the hydrogen line was an excellent frequency on which to listen. This is a natural frequency known to all communicative races and one on which a great many radio astronomers are busily observing. Because of doppler shifts we would still have to search a 3-MHz band but this is only one-thousandth the task posed by the entire 3-GHz band. To avoid the noise associated with the hydrogen line itself, others have suggested searching at one-half or twice the line frequency.

In the last few years many other spectral lines have been discovered, so the hydrogen line has lost its unique status. Furthermore, there are good reasons not to select a single frequency, and particularly a spectral line, for both transmission and reception. If we do both simultaneously, we jam ourselves. If we transmit half the time and listen the other half, we halve the probability of being detected and double the search time. Further, we interfere with radio astronomy work on the spectral line itself. Obviously we cannot all choose one frequency for transmission and another for reception. What then is the best strategy?

To eliminate self-jamming, what we are seeking is a naturally defined *band* rather than a single frequency. The band should be wide enough to provide adequate frequency separation between transmission and reception, and narrow enough to permit simultaneous search over at least half the band. It should contain no spectral lines to interfere with reception or to be interfered with by beacons. The band should be at the low end of the microwave window but not below the hydrogen line where observations are made of the red-shifted hydrogen lines of other galaxies.

There is one band that seems to meet all requirements. It is the band between the hydrogen line (1,420 MHz) and the lowest of the hydroxyl lines (1,662 MHz). It provides 242 MHz of clear spectrum at very nearly the quietest part of the microwave window. (On planets with more water vapor and oxygen it would be the quietest part.) What more poetic place could we find for water-based life to seek its kind than the band defined by the two disassociation products of water itself. Let us meet, as different species have always met, at the water hole!

In addition to using a natural frequency, having high power, and operating continuously, beacons should be information bearing and be circularly polarized. The information would contain among other things instructions as to how to respond. The required modulation could have a number of easily detectable forms but should not degrade the strong monochromatic carrier strength. Circular polarization stays circular, whereas linear polarization suffers Faraday rotation in the interstellar medium and has no preferred orientation to begin with. With circular polarization the receiver need only examine the two orthogonal possibilities (right and left) to be sure of receiving the signal with zero polarization loss. Knowing that circular polarization makes the reception task simpler, all sending races will choose it. Incidentally, the choice of circular polarization allows the sender to communicate what he means by "left" and "right." All that is needed is to send a diagram of the arriving wave. The arriving wave itself resolves the ambiguity. The medium becomes half of the message.

The Cyclops System

From the preceding calculations and discussions it appears likely that the detection of coherent signals of intelligent origin will require a collecting area on the order of 7 to 20 km², corresponding to a clear circular aperture 3 to 5 km in diameter. The only practical method of realizing the large or collecting area seems to be with phased arrays of smaller antenna elements. Our estimates of the density of intelligent life in the universe and of the power levels this life might radiate as leakage or in beacons are so uncertain that this antenna area might be one-tenth or ten times the figures given above. This uncertainty itself is a strong argument for a phased array which can be expanded only to the extent necessary as the search proceeds.

Much of the Cyclops study was devoted to finding engineering solutions to the problems of providing low-noise, remotely tunable receivers, of phasing the local oscillators, of transporting the IF signals back to the central processing station without selective attenuation or delay variations, of providing the delays and phase shifts needed to steer the array, and of controlling and monitoring the array operation. These solutions are discussed in the Cyclops report and contain many novel ideas that may be of value in the design of other arrays.

Here we need only say that *it now appears practical to build a closely spaced array having an equivalent clear aperture several kilometers in diameter, tunable over the entire microwave window.* If desired, the array can also be subdivided into many subarrays each used for a different purpose and at different frequencies, at the same time. It also appears possible to image the radio sky for mapping purposes and for picturing remote sources on a real-time basis. Unless novel approaches can be invented, the cost of the entire system is dominated by the cost of the antenna structure and would amount to about six billion dollars for an array 10 km in diameter with an equivalent clear aperture diameter of 3 km. Future cost-saving ideas may reduce this figure considerably. Nevertheless the figure quoted does not seem unreasonable for a project having the magnitude of the search for extraterrestrial life.

As conceived, the Cyclops system would carry out the search largely automatically. Service and maintenance would be needed, but the search of whole classes of stars would proceed without human control or monitoring. The coordinates of target stars, stored in the data file, would be read in sequence, and the observation of each would be made automatically. Only if an anomaly was discovered, would human attention be invited. This degree of automation is required to eliminate, or greatly reduce, the psychological depression of devoting years of daily effort with negative results.

It is the signal-processing system that most specifically qualifies the Cyclops system for its primary search mission. The complete array prior to the data-processing system delivers two IF signals, each representing the output of the entire array for the received band. Each IF signal has a bandwidth of 100 MHz, and the two signals represent the two orthogonal polarizations to be analyzed. The data-processing system, in 1,000 sec of observing time, will detect any coherent signal, even a doppler drifting one, whose power is not more than 90 db below the noise power in the 100 MHz bands.

The first step is to transform the received signals (amplitude vs. time) so as to obtain successive samples of their power spectrum (energy vs. frequency). This converts the nearly sinusoidal waveform of any coherent, nearly monochromatic signal into a sharp peak in the frequency domain. In any given sample of the power spectrum, this peak may not be conspicuous compared with other noise peaks. However, if a large number of samples of the power spectrum were displayed in a raster, the continued presence of the spike from the coherent signal would produce a clearly visible pattern. This is illustrated in Fig. 9.4. (The photograph is actually a representation of the signal intensity vs. time in the outputs of 50 receivers, tuned to adjacent frequency bands. Each line in the photograph is the output of a particular receiver. The signal, which appears as the slanting line, is a single pulse from a pulsar. The pulse is received at different times in each receiver because of the delay dispersion of the interstellar medium.) For our purposes we can think of each line as a segment of one sample of the power spectrum, that is, a representation of intensity vs. frequency. The slanting line now represents a coherent signal drifting slowly in frequency.

By covering all but one line in the photograph it is obvious that the signal would not be detected in any one sample of the power spectrum. Yet it is clearly visible in the ensemble. We can recognize the straight-line pattern electronically by adding all the power-spectrum samples in such a way that the peaks from the signal accumulate directly in the sum.

Figure 9.4   The pattern produced by a drifting coherent signal.

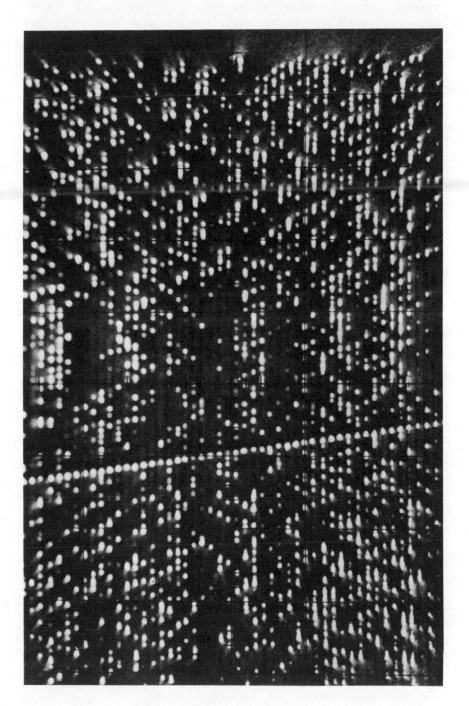

Hence the second step is to add the successive samples of the power spectra under a variety of frequency offsets between adjacent samples to allow for any reasonable drift rate the signal may have had during the observing period. In one of these additions, the one that matches the drift rate, the signal peaks will add to form a spike that is clearly above the noise level. There the final step is to determine, by simple threshold circuits, whether such a spike exists. If so, the observation of the star is resumed and if on the next analysis the spike is still present (in the frequency range where it should have drifted), a third test is made with the array directed a few beamwidths off the star. If the spike disappears and then reappears when the star is re-acquired, an alarm is sounded. Otherwise the star is merely designated as a candidate for further examination.

The breakthrough that makes this data-processing technique possible is the optical spectrum analyzer. The two-dimensional Fourier transforming capabilities of lenses using coherent light are well known. Less well known is a technique of bringing this full capability to bear on a one-dimensional signal.[2,3] The signal is first recorded in raster form on a moving strip of photographic film. (A constant bias is added to prevent negative amplitudes.) After processing, the film passes through a gate where it is illuminated by coherent light as shown in Fig. 9.5. If the film in the gate is at the back focal plane of a lens, the Fourier transform of the amplitude distribution on the film will be formed at the front focal plane. The *intensity* of this transform is the desired power spectrum. *The power spectrum is also displayed in raster form.* Each line of the power spectrum raster displays a portion of the spectrum whose width is equal to the scanning frequency used in the recording process. The frequency resolution along the raster line is equal to the reciprocal of the time represented by the total recorded signal segment in the gate. Thus, if a 1-MHz signal has been recorded with a 1-kHz sweep frequency and if 1,000 lines of the recorded raster are in the gate, this represents one second of received signal. The power spectrum will also consist of 1,000 lines, each of which represents a 1-kHz band in the spectrum displayed with 1-Hz resolution. Such an analyzer would be said to have a time bandwidth product of 10^6 and is a currently available piece of equipment.

Analyzers with time bandwidth products of 10^7 are believed to be within the present state of the art. Twenty such analyzers would be needed to analyze the two 100 MHz cyclops IF bands into 1-Hz "channels"; two hundred would be needed for a resolution of 0.1 Hz. No other known analyzer approaches the real-time capability of the optical analyzer.

The IF signals are first separated into bands of appropriate width for the analyzers used and then heterodyned down to a convenient baseband.

[2] C. E. Thomas. "Optical spectrum analysis of large space bandwidth signal," *Applied Optics,* **5**, (1966) 1782.
[3] R. Markevitch, "Optical processing of wideband signals," Third annual recording symposium, Rome Air Development Center, April 1969.

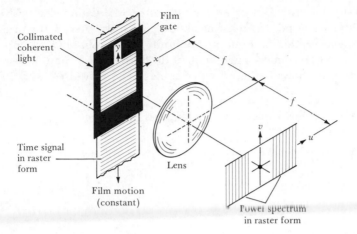

Figure 9.5 Optical spectrum analyzer.

The total film usage is small and depends only upon the bandwidth analyzed and the resolving power of the film. If ten times as many channels each one-tenth as wide are desired, ten times as many analyzers are used and the film runs at one-tenth the rate in each. The film area needed per second is

$$\dot{A} = \frac{2B}{k^2}, \tag{9.24}$$

where B is the bandwidth of each IF signal, and k is the number of resolvable lines per unit distance. Taking $B = 10^8$Hz and $k = 3,000/$cm, we find that $\dot{A} = 20$ cm$^2/$sec.

In the cyclops system the power spectra are imaged onto vidicon camera tubes, where they are scanned and the video signals are recorded onto flying head magnetic disks. As many as a hundred or more complete power spectra representing successive frames of film in the gates are recorded for each observation.

The power spectra are then played back simultaneously and added together with various amounts of relative delay between successive spectra. This is accomplished by sending all the spectra down video delay lines and scanning the signals from taps disposed in slanting rows across the columns of delay lines.

Suppose that the two 100-MHz IF signals are recorded for 1,000 sec for each observation, thereby obtaining 100 samples of the complete power spectrum with 0.1-Hz resolution. These may be added in several hundred ways to accurately match all possible drift rates, but since additions with almost the same slope produce correlated signals, only 100 independent signals are obtained in the adding process each having 2×10^9 Nyquist intervals per observation. Thus 4×10^{11} independent threshold tests are made per star observation. To keep the probability of a false alarm to 10 percent for the entire observation, the probability of a false alarm per test must be about 2.5×10^{-13}. With the thresholds set to achieve this false alarm immunity, the probability of missing a signal at the reference range limit (0 db SNR) is less than 50 percent.

Figure 9.6 shows the input signal-to-noise ratio required, as a function

Figure 9.6 Signal-to-noise ratios required for detection. (False-alarm probability per datum is 10^{-10}.)

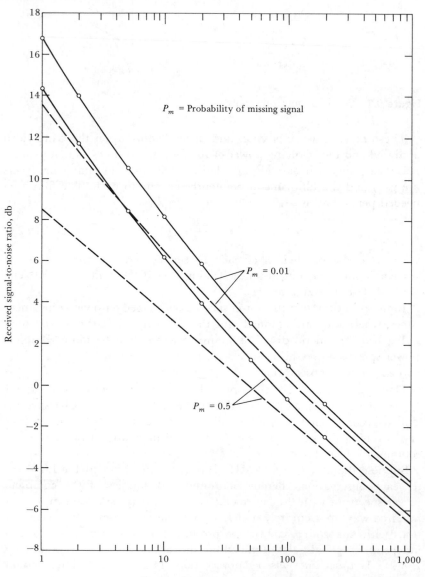

Number of independent samples averaged

of the number of samples added, to have a 50 percent and a 1 percent chance of missing the signal when the false alarm probability is 10^{-12} per test. (Going to 10^{-13} raises the required SNR by about 0.1 db.) The dashed lines are the curves obtained if gaussian distributions are assumed rather than the actual statistics of the noise and noise plus signal from a square-law detector.

Cyclops as a Beacon

Although the Cyclops system was conceived as a receiving search system nothing prevents its being used in the transmission mode as a very high-capability interplanetary radar or as a cluster of beacons. All that is needed is to equip each antenna with a small transmitter. With 10 kW per antenna and 1,000 antennas, the total transmitted power of 10 megawatts could then be directed, with all transmitters phased, as a single beam having an effective radiated power of about 2×10^{16} watts or as 1,000 separate beams each with 2.5×10^{10} watts of effective radiated power.

The arguments given earlier for making *long range* (1,000 light-years) beacons omnidirectional do not apply for ranges under 100 light-years. For these ranges the number of target stars matches the number of antennas in our "receiving" array. Thus we might well consider transmitting to these stars if the first search has detected no signals from them. The details of an optimum search strategy remain to be worked out.

A Comparison of the Cyclops and Ozma Systems

The only significant attempt ever made in the United States to detect interstellar signals of intelligent origin was made in April and June of 1960 when Frank Drake and his associates listened for 400 hr for evidence of artifact signals from two stars in the 10- to 11-light-year range: Tau Ceti and Epsilon Eridani. Project Ozma, as it was called, found no signals except for a couple of very exciting false alarms.

We present in Table 9.3 the significant parameters of the Ozma system and the "bogey" Cyclops system of 3.16 kilowatts clear aperture.

Table 9.3

Parameter	Symbol	Ozma	Cyclops
Antenna diameter	d	26m	3,160m
Antenna efficiency	η	0.5	0.8
System noise temperature	T	350°K	20°K
Resolved bandwidth	Δf	100Hz	0.1Hz
Integrating time	τ	100sec	1,000sec
Instantaneous bandwidth	B	100Hz	200MHz

Taking the sensitivity to be proportional to $(\eta_d^2/T)\,(\tau/\Delta f)^{1/2}$ we see that the sensitivity *ratio* of the Cyclops to the Ozma System is

$$S_{c/o} = \frac{0.8}{0.5} \cdot \frac{3{,}160^2}{26^2} \cdot \frac{350}{20} \left(\frac{1{,}000}{100} \cdot \frac{100}{0.1} \right)^{1/2}$$

$$= 4 \times 10^6$$

Since the range limit for a given radiated signal varies as the square root of the sensitivity and since the volume that can be searched is proportional to the cube of range, we see that the range and volume ratios of Cyclops over Ozma are

$$R_{c/o} = 6{,}000 \quad V_{c/o} = 2.6 \times 10^{11}$$

The target stars of Ozma, namely Tau Ceti and Epsilon Eridani, are about 10 to 11 light-years from us. Any signal that the Ozma system could have detected at this range could be detected at 60,000 light-years by Cyclops, or could be 1/4,000 as strong and be detected at 1,000 light-years.

In addition, the proposed Cyclops system searches 2 million times as broad a band in ten times the observation time. Its spectrum search rate is thus 200,000 times faster.

These comparisons are made not to disparage Ozma, but to build faith in Cyclops. Ozma cost very little and was a laudable undertaking. But the power of the Cyclops search system is so much greater that we should completely discount the negative results of Ozma. The Tau Cetacians or Epsilon Eridanians would have to have been irradiating us with an effective power of about 10^{12} watts to have caused a noticeable wiggle of the pens of Ozma's recorders. 250 kW would alert Cyclops.

Conclusion

The search phase is the most difficult aspect of interstellar communication. Once contact has been established, the transmission of information can take place with modest powers and antenna sizes. Out of earlier work and the Cyclops study a system is beginning to emerge that shows promise of being able to accomplish the search task. Significant progress has been made in one of the most formidable of the problems: that of searching the spectrum for narrow band signals. There now appear to be no technological barriers that prevent us from building receivers with square miles of collecting area capable of detecting beacon signals from as far away as 1,000 light-years, and of detecting weaker beacons and leakage signals from distances up to about 100 light-years.

The search phase may discover signals in a very short time or it may take centuries of effort. At our present state of knowledge we simply cannot say what the time scale might be. If the time turns out to be short, it will be because other races have engaged in a sending effort for very long times. For all we know, this could be true. Perhaps a vast heritage of galactic knowledge and wisdom is being passed on in this way to younger civilizations as they mature. If so, then making interstellar contact will profoundly affect the destiny of the human race.

Where Does All This Lead?

The previous chapters have discussed many different facets of the interstellar communication problem. In this final chapter Philip Morrison looks at many of these issues from a cosmic perspective.

Morrison, in collaboration with Giuseppe Cocconi, was responsible for generating scientific interest in the interstellar communication problem by a letter to *Nature* suggesting that extraterrestrial civilizations might be attempting to communicate with us on the 21-cm-wavelength band. It is fitting that the final overview of the problem be presented by him.

10

Conclusion: Entropy, Life, and Communication

Philip Morrison   *Massachusetts Institute of Technology*

History

This discussion is not a detailed summary of the other chapters, rather, it is a summary of the issues involved: it presents the author's opinions of the way in which the various issues will be joined and of the disputes and uncertainties which remain. In this informal discussion we shall cover, in an episodic way, now one point, now another. It was Parmenides who said: "Now we leave the realm of the shapely, well-rounded truth to discuss mere mortal opinion." I shall try to make a distinction between shapely, well-rounded truth and mortal opinion. This does not mean that I do not now believe in my opinions. It means that I recognize that many of them are probably wrong. (Of course, that is often true of the "well-rounded truth," too, in this imperfect world.)

About ten years ago interest began in publications in the field of interstellar communication, apart from the fruitful area of science fiction. The pioneer work was that of Frank Drake and the late Professor Otto Struve at Green Bank, in the early days when they actually used the 21-cm telescope to listen to a few of the nearby stars. They established what up until that time was not known at all: that not every star of solar type was the origin of a very powerful 21-cm modulated signal.

It was not until 1960 that we could be sure that such was not the case. You could easily imagine a strong signal coming from any of these sources, for nobody had listened before 1960. It has turned out that the problem of communication is not simple, which is to be expected, yet it is something of a pity. Since that time, the ratio of papers to experimental results —except for those of Drake and Struve—is unfortunately high. It is almost time to stop theory. Later in the discussion we shall consider what the next practical steps might be. The Soviet scientists are also very much interested in the subject and have published two volumes of symposium reports (available in translation).

Travel or Telecommunications?

Let us approach some of the points that have been raised both generally and also specifically in the preceding chapters. First, let us consider travel. It is important to recognize that the official title of this collection uses the term "communication." Travel is much more than communication. It is the transport, not merely of a little free energy, informational entropy or free energy, from place to place, but rather the transport of massive material structures. Now, the transport of material structures over galactic distances with a speed approximately the speed of light is probably beyond all accomplishment. In my opinion we shall never do it. It is dangerous to speak that way. One of the consequences of such a statement is that perhaps it will challenge somebody in the distant future to do it. I still don't think it will be accomplished. The fuel problem for relativistic rocket travel is overwhelming. The other two approaches that can be taken are both very ingenious. But one of them will fail on economic grounds and the other one, which concerns stability, will fail on more technical grounds.

The first proposal is actually a mix of travel and communication. In his ingenious proposal Bracewell suggests that the best way in which communication could be made would be to establish relay stations by distant probes, which could carry themselves bodily to the near neighborhood of the potential partner in the communication channel, prepared to sit there a long time. Once the potential partner made his first feeble efforts to begin communication, these drones, being close at hand, could pick up that fact, respond, and make it known back at base. It is a good idea. It avoids the great trouble of having a beam enduring for a very long time directed towards a station which almost never responds, a situation which is not unfamiliar in communications point-to-point even on the earth. It is certainly true that his is a continuous early warning system; that is, it would provide the earliest notice of the development of potential partners that we could imagine. For it would not be necessary for the new partner to have equipment enabling it to communicate with a distant source; the new planetary partner would only need to communicate with some device in its near stellar neighborhood. A probe traveling in a solar orbit is the typical model which Bracewell describes. Against this system there is one very heavy criticism: interstellar communication or travel over perhaps a couple of thousand light-years, even for these probes which are machine-carrying but not man-carrying, is a task which is hard to carry out at high relativistic velocities. Therefore it seems that the transit time could be estimated to be many tens of thousands of years, traveling at a tenth of the speed of light; one estimate, roughly speaking, might be 10^5 yr. It is not reasonable to make a probe which will be in transit for 100,000 yr and have a duty life not larger than that travel time; that is certainly very uneconomical. Therefore we must imagine a probe which is prepared to sit around and listen on the order of a million years. Such a probe is very useful, but I regard it as a very difficult object to manufacture. Under no circumstance can I imagine that such a probe, which must go to a remote point and maintain itself, will appeal to a technologist, however great his technological strength, when he can do the same thing at home without sending the material by maintaining a transmitter within his own backyard, so to speak. Surely a probe sent out so long ago would become obsolete. I do not think the problem of putting a lot of power into a CW signal or of performing a similar procedure is nearly as serious as the maintenance of an up-to-date remote station for a million years. The scheme is highly ingenious but on economic grounds, so to speak, it does not seem plausible.

Perhaps one of the original stimuli for Professor Bracewell's ideas was the existence of records of semi-regular radio-frequency echoes with very long delay repeating terrestrial transmissions. In the last 30 or 40 yr, it has been noticed that under very special circumstances, radio signals emitted from normal point-to-point radio transmitters returned only somewhat dispersively true echoes, with delay times as large as several tens of seconds. This is a very long time to be accounted for by the ionosphere; it therefore suggests some remote trip and return. It perhaps led Professor Bracewell to think for a moment that an echo device in distant orbit was causing it. (However, he views it only as an analogy.) In the context of having sent out an extremely expensive probe for 1 million yr of duty, with 100,000 yr of transit time, a probe should, however, make itself known in an unambiguous way. Since it is almost as easy to make an unambiguous signal as it is to make an ambiguous signal, you might as well put in the extra circuit card. If I say, "Hello," and the echo says, "Hello," which is the case in point, that does not impress me as being a very powerful mechanism. It turns out, as we all know for acoustic echoes, that some simple natural structures can do that, a cliff, for instance. I do not know what would cause an echo in the ionosphere or in the earth's magnetic wake. (That is the real problem; we are not trying to solve that geophysical problem here, but let us stress that a simple echo is not an information-filled response.) If I said, "Hello," and the machine heard it, a really good machine would say, "Hello, how are you?" Then I would be sure that there was something out there, not a simple system, but something that has a very intricate pattern of stored free energy, of the kind not seen elsewhere in nature. That is the key. Nobody will make a system without a really clear way of disclosing itself. Since we have not seen that, we can assume that a probe does not exist in this instance. That does not mean they will not exist someday or somewhere, although on the economic grounds described above, that is not likely. That kind of "travel," which I think is the cleverest travel proposal yet, does not, therefore, convince me.

The second kind of travel is real interstellar travel, where people, intelligent machines, or whatever you like, go out to colonize. You travel in space as Magellan circumnavigated the world. I do not think this will ever happen. It is very difficult to travel in space. We will not repeat the arguments which have been made frequently in print. A very attractive proposal by Dr. Bussard would circumvent the problem of having to carry the enormous supply of fuel required for a free rocket by the exponential penalty for payload to fuel-load ratio. You must use not a prefueled rocket at all, but a ram jet, which takes the interstellar gas as a source of its nuclear energy (see Fig. 10.1). This is a very ingenious proposal; in my view, it is the only travel proposal which has any chance of working. During the last year or so we examined this system and we came to several conclusions. First, we asked what would such objects look like in space? We had to admit that there was no clear evidence that they were not there! They do not make such conspicuous shocks that we would necessarily see them. They are worth thinking about, even worth looking for. It is an attractive exercise for anybody who has a favorite kind of astronomy, radio, infrared, gamma ray, gravitational, whatever it is, to see if he can find any possible observable effect of such a vehicle anywhere in the galaxy and to try to locate it. But a much more serious point is that these ships suffer very heavily from profound radiative losses connected with their own flow instability. They have to spend a great deal of energy. Let us consider a ram jet: now, we do not propose to build one, so we are not going to describe the mechanism. All we are going to do is solve the conservation laws. We are sure of those; that was Bussard's great idea. We can design a device in terms of the conservation laws; let the technology of the future find out how to build them. Assume that we have some tremendous machine which makes nuclear energy out of the deuteron-deuteron reaction or maybe even out of the proton-proton reaction itself. One hundred miles away from it or perhaps ten miles away, we put a ten-ton payload for a man

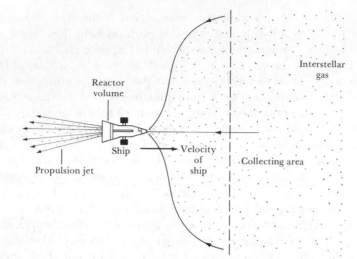

Figure 10.1   An impression of the interstellar ram jet. The challenge is in the necessity of collecting gas from a disc which measures thousands of square dilometers and which must then be channeled into a reactor of normal engineering size. The affair is plainly unstable. Does the flow, even if stabilized, give rise to inevitable radiation losses by the electrons and ions of the plasma?

and his computer, and then this machine has to carry a great collecting disk. All the fuel to make it go relativistically, and to beat the drag, is coming in from the scarce atoms of the interstellar gases, a most excellent vacuum; therefore the disk diameter must have a radius of at least 10 or 20 km. Now, we do not know how to make such devices, but that is not the point. You must imagine some kind of widespread magnetic field that this device generates which causes all the ions to flow through the engine that does the nuclear job. Of course, that's the trouble; it is a terribly unstable situation. What we seem to be able to show with a fair degree of generality is that even allowing for a solution to all problems of stability from plasma-flow considerations, the needed change in velocity for electrons (electrons must be separated at most by the Debye distance from the protons) requires sufficient acceleration so that the electrons radiate away more energy than is being gained, if you use only rare deuterium. Probably they radiate nearly as much energy as you are getting even if you can use the protons. It looks as though it is a marginal system, even examined in this "in principle" way. I believe that this device also will not work; if it does not work, then we will not find any way to travel very far in space.

It seems, then, that men will never, or hardly ever, travel great distances in space, that is, except roughly on a set of measure zero. However, once there is really interstellar communication, it may be followed by a ceremonial interstellar voyage of some special kind, which will not be taken for the sake of the information gained, or the chances for trade (clearly negligible compared to the expense), but simply to be able to do it, for one special case, where there is a known destination. That's possible, one can imagine it being done—but it is very unlikely as a search procedure. Therefore I am led to the conclusion, unchanged after 10-yr time, that indeed we do have to think mainly about communication and to ask where in the frequency spectrum. That's the first decision; then we can return to the old discussions.

What Message Channel?

It seems clear, and Drake, Bracewell, and Oliver have reinforced this conclusion, that the gigahertz region, where the noise is minimal, is still the region of choice; the noise rises toward lower frequencies because of galactic nonthermal radio sources, and it rises toward higher frequencies both because of quantum noise and because of the possible effects of stellar interference, which is very difficult to block out from the hot surfaces of stars. Whether the best choice is exactly 1,430 megacycles, the hydrogen hyperfine line, one may well doubt. It is still a possibility, since it comes sufficiently close to the region of the best place to be attractive; perhaps its uniqueness is somewhat disturbed by the many rf-lines which have since been found. These throw a certain doubt on the uniqueness of this frequency choice (between H and OH, as described by Oliver). This decimeter region remains attractive.

There is, of course, the possibility which has been plain for a long time, that acquisition and message transfer are two different stages of long-distance communication. Again, everyone who is involved with point-to-point communication knows that. If you listen on your HF receiver to point-to-point signal channels, amateur or commercial, you will find that much of the channel time is occupied with call signs and strings of CQs. These are acquisition procedures. The local radiotelegraph stations, marine and point-to-point, keep sending CQ CQ CQ, hoping for somebody to break in, keeping a channel open so that they can be picked up. There is no information content except for the existence and activity of the station itself. It is rare when a reply frequency or some simple data like that is also mentioned. This signal design is typical because acquisition is always a special procedure. You could, for example, change your frequency once you make contact. You could widen bands. You could increase power. You could try many arrangements, because now you are spending your energies in a time domain which has really paid off, whereas the time domain of sending while nobody listens to you is always a frustrating time to operate. It is quite natural to divide these two tasks. I believe we need more work on the problem of distinguishing the acquisition channel from the message-transfer channel. Acquisition might be done by very simple means by exploiting natural energies, for example, by placing a source of atomic contamination on the surface of a star; the resultant strange but characteristic optical emission line might prompt someone to say, "A star that emits a narrow line of gadolinium, what can that be?" This fails by a little bit the test of uniqueness. Recall that when the OH line, the formaldehyde rf line, and many other lines not in equilibrium were found, there was first a feeling that this could be a message of some sort. It was then established that this was not the case, but that a new physical phenomenon was responsible. We cannot use quite such simple-minded methods. A little more information content is needed than just the existence of a line of something rather rare or a very regular pulse, because if you look for a sufficiently long time, you are bound to find very rare things just in the course of events. This is a very valuable lesson. Although *rare* things occur *rarely,* it is also true that rare things *occur* rarely. That sentence has much meaning.

After ten years I remain pretty much in the same place. I come back to the old conjectures with which the names of Drake and the rest have been so long connected: those famous formulae that have to do with how many stars are sunlike, stable, and not multiple; how many have adequate domains of warmth; what fraction of them have planets; what numbers of planets we can expect; what is the probability of the evolution of life; and what is the number of communicative species (by which we mean species able to make signals across interstellar space—our narrow, operational, utilitarian definition)? (These conjectures do not ask whether the effort is worthwhile; they just ask *can we do it?* For if we cannot do it, even if it is worthwhile, we will not be detected.)

Who Is Sending?

The most uncertain number of all (to which Aronoff addressed himself in part in Ch. 6) is the length of time this phase of communication might endure. First of all, consider the famous formula:

$$N_c = N_* \cdot P_\Theta \cdot P_E \cdot n_E \cdot \beta \cdot n_{sp} \cdot T_c / T$$

where N_c is the number of communicative sites, the number of origins of signals; it is given by N_*, the total number of stars, times the probability that a star will be like the sun P_Θ (this term will be defined a little more closely), times P_E, the probability that the sun will have an earth-like planet, times n_E, the number of such earth-like planets you may expect (using an average of some sort to indicate how uncertain these matters are), times β (which stands for Bios, life), the probability that it will have living forms, times n_{sp}, the number of communicative species evolved in such a system, times that fraction of the time which is the duty cycle, that is, how long the system is turned on and communicating, which is the time it lasts while communicating, T_c, divided by the total time T, which is some measure of the total evolutionary age of a planet or a star, say, 5 billion years conventionally. Factors of two, of course, do not concern us. That is the usual state of affairs, and we have heard the numbers. They have been discussed at length with all the latest information by Professor Cameron, especially on the astronomical side, the first half of the factor. Perhaps I shall be a little more conservative than some. The first four factors are the astronomical side, the next one or two are the biological side, and the last one or two are even more complicated, the social side. We know least of all about the most complex problem, though it is closest at hand.

The number of stars in our own Milky Way galaxy can be taken to be about 10^{11}. What is the probability of sunlike stars? We have to define what we mean. Stars that are too bright must be excluded because they do not last sufficiently long for the slow processes of biological evolution to take place; this means that periodically all stars hotter than spectral type F are eliminated from consideration. The stars must be stable enough and free of multiples.

Science fiction writers, even the less imaginative ones, have repeatedly pointed out to us that there are many other possibilities. In my opinion, the most extreme and the most brilliant possibility is that suggested by Fred Hoyle who conceived of a form whose size was not measured on terrestrial dimensions at all, but on solar-system dimensions. Its "chemistry" was not atomic chemistry at all, but was plasma physics. Its information store was electromagnetic, perhaps Alfvenic in nature. It is hard to show where Fred Hoyle is wrong. He has his Black Cloud tell us there was no first one, but he does not say how the others were made from the first one. You just imagine these beasts can do anything they like. As long as you are not limited by historic processes, you don't have serious problems inventing all sorts of objects, although they occupy states which matter may never reach. But I would like to include all those beasts, the silicon beasts, the plasma beasts, beasts that live contentedly near triple stars and take advantage of the three different colors, and whatever else you can invent. Imagine whatever you want, there is plenty of possibility for invention; all of those I would like to take as a bonus. I don't know how big that bonus is; it may be zero (I think it probably is), but I am sure it is not negative! If it is not smaller than zero, then all it does is add to the estimates that I am going to produce. We should try to work on the minimum estimates we can make.

Therefore, let us start with the system where we live, although we do not understand it very well; let us see how much we can imagine, deviating from that only conservatively, letting everything work more or less as it works now. I admit this is not an absolutely unambiguous prescription, not of mathematical rigor, because what do I mean by more or less? Do I require the shape of the Mediterranean Sea to be the same, Italy to be shaped like a boot, or else our species would not have developed Greek civilization? If you take such a position, then of course you will easily prove yourself utterly unique. In the astronomical domain we can clearly spread the variables a little bit.

Let us consider a sun which lasts billions of years, which does not fluctuate because of strong flares, which does not evolve strongly in the time in question, and which is not a multiple system so that the insulation on any planet remains constant. Then calculate the expected number of planets warmed by the star. Allow that to fit into the galactic distribution of stars. As you know, everybody comes out with about the same optimistic answer; that the number of sunlike sites in our galaxy is of the order of 10^8. Let us simply say that it surely lies between 1—which is us—and 10^8. That is about the most we can honestly say for sure. The experts on stellar evolution and planetary formation would say more, but I emphasize we do not yet have a sure theory of these processes. In the absence of a clear theory, we can only make guesses and estimates. Making those estimates, assuming we are not unique, there will be something up to 10^8 planets of our kind.

Do we have any evidence of these planets? The evidence that we can obtain consists of the planetary perturbations of the motion of the central star, due to the presence, not of tiny earth-like planets, but of massive planets like Jupiter. This phenomenon has been observed certainly on one, and probably on two of the two or three closest stars; that is not a proof, because the planets "seen" by Van de Kamp and others near Barnard's Star are not of the terrestrial sort. The star itself is probably too dwarfish to be hospitable, but still it serves notice to us that where we could look for such small perturbations, we did indeed find them. That suggests that maybe the planets are not so rare as one might think. In the absence of knowing about planets, our arguments are only indirect. The disappearance of angular momentum proves that something went out of the star, but it does not prove at all that planets were formed. It might have simply been dusty rings, or something too fragmented to be hospitable for the development of the gas and liquid chemistry that we probably need. This is one of the great uncertainties. Obviously, we are anxious to learn about planetary formation for other reasons; all we can say is that this is a very important subject that we would like to study carefully. If we take an optimistic view, we have a number something like 0.1 to 1 for $P_E n_E$. That leads me to the biological side.

What Is Life?

First, what do we mean by life? How can we define it in some way? We know what we mean operationally; it must build radio transmitters. Although that is the next step in the discussion, we must ask the easier question first. There is no hope for finding a definition of life if we cannot argue from structure to behavior. That is to say, it is not useful to try to define it by saying that life is something that reproduces, or that it has this, that, or the other property, because no substantive definition is gained that way. A rock is not something that has this and that property; a rock is a collection of minerals, coarse-grained or fine-grained, and so on. It is described in terms of structure. That is what a rock is; that alone enables us to discuss it genetically, because it tells which elements it is composed of, and how they interact.

At first this seems to be a very difficult problem. I came to what I think is a happy, simple-minded solution to the problem of describing things in terms of structure instead of in terms of behavior. Let us consider an example: If we take a rabbit and make just one small structural change in this poor rabbit, for example, cutting some crucial nerve or cutting the aorta, with that small change in structure, the rabbit quickly dies. There is hardly a microscopic change, except locally. The whole atomic structure of the rabbit, barring 1% somewhere, is the same. This, of course, is the problem of an organism, that of organization. It appears as a great philosophical barrier; it has caused a large number of books to be written. But from a naive, oversimplistic physicist's point of view, we can circumvent that problem nicely. I claim that the discussion of what is living and what is not living cannot usefully be put in terms of individual organisms. I will not be able to put it in that way. That is not new physics. We can not say what the temperature of a gas is by looking at one gas molecule. It just does not make sense, because any molecular velocity may be present in a gas of a given temperature. Just as the idea of temperature is a statistical concept, I believe also that the idea of a living organism is a statistical concept. The whole idea of a species, and so on, is statistical. (You could devise special cases of species in which only one individual now exists, but these are really quibbles.)

Let us consider an example which frees us to talk about the matter in a structural way, which we could not do before. What is a clock—a working clock? Of course, there are many kinds of organisms. Let us suppose that I take the clock apart, and I throw the gears on the table. It is certainly no longer a clock. But suppose that I take that complete clock and I place a small grain of dust between two pinions in the escapement where there is not much force. The clock stops. Now, it is still a clock, in the sense that it should still have that label, although you could not easily sell it in a clock store. They would say it does not behave like a clock, and therefore it is not a clock. So you are led to the trap of thinking in terms of behavior. Now you are in trouble because after all, the structure of that clock is all but 100% clock structure, barring a tiny addition of a few parts per million. What is wrong with this view? Let us consider the following: suppose that we take millions of such clocks, a whole ensemble of nearly identical units, all good clocks with little specks of dust added. The dust particles are added at random. Almost all of those clocks work perfectly! A clock is, in fact, not really defined up to a random little grain of dust.

Consider the clock on your shelf which is working beautifully. Even excellent housekeepers will leave several grains of dust on it here and there which do not prevent it from being a perfectly good working clock. How does that differ from the case of the clock which has the one grain of dust in the crucial part? One might say, it differs in that the one crucial grain of dust spoils the function. But we seek to escape from that circular definition. So we say that it does not really differ at all. When we say a working clock, we describe a normal brass clock by giving the position of all the gears, the composition and size, the coordinate metallurgy of the whole clock. What we are doing is not describing an individual clock, rather we give a range of dimensions for each gear or pinion, say all the metal-bearing points in three-dimensional space on a 0.1-mm lattice, each given a number, and with the statement whether it is brass or bakelite or air at any point. We can describe the whole clock roughly in a vast multicoordinate phase space. Then we can say, there is a clock, and next to it, there is another clock, and next to it, another one, and next to it, another clock—a whole range of clocks. It is true that some of them are *not* clocks, in the sense that a few of them do not work. But there are also clocks that do work, and the average structure in this neighborhood is a working clock. We thus relate structure and behavior, not in every individual case but *on the average*. This is indeed more comfortable to common sense: if somebody offers you cheap a perfectly good clock which does not work, because only one thing is keeping it from working, one little speck, you buy it. For you can remove that grain of dust and it will keep time well. The point is that it was much closer to being a working clock than just any pile of gears on the shelf, and much closer than a brass ingot, and far closer than a piece of copper ore and a piece of zinc ore. Structure, then, does contain all behavior in exactly that sense, but it must be regarded statistically. The more complicated it is, the more you have to average to say what the structure to cause the behavior is. For many behaviors can follow from a single approximate structure.

The same thing is true of a computer. For example, if you bought a derelict computer and plugged it in and nothing happened, but if it is a perfectly good computer, there may be something small wrong with it. If you happened to know that there was a break in the 110-volt cord, you have really got a bargain. A nonworking computer can become a working one with one simple change, in the same way that the passage from life to death or from death to life occurs. To put it dramatically, if I try to kill a man simply by stopping his heart, we now recognize that in some sense he is not dead; we may be able to reverse all the damages because in the structural near neighborhood of that man, there were plenty of phase points where he was perfectly well. You could have cut many organs just as deeply as that, and it would not have hurt him to the same extent. It is just the unhappy chance that your scalpel hit the neighborhood of the cardiac nerves that caused the trouble! If you define life statistically, it covers even this situation. The same is true of an aircraft or of any other "organism" as far as I can tell. We can describe what an airplane is, not only by saying it has to fly, but by saying what its structure is in some statistical way. Most of the structures in its structural neighborhood do fly.

Therefore it appears that we will be able to characterize living things on a quantitative scale which will describe them structurally. The key is the free-energy content, the usual $E-TS$ of a system which is self-reproducing. In the self-reproducing process, a good deal of free energy remains stored within the system, as free energy of structure, informational free energy. The fraction so stored is a continuous parameter. That continuum was traversed in the early stages of the origin of life, when molecular structures that were not quite complicated enough to have a very large storage of free energy of structure grew into a situation where they were definitely disappearing. Now some self-reproducing structures are merely logical models, like those of Penrose or Morowitz; they are like little blocks or little toy trains made so cleverly that if you placed one in an environment full of rather similar things it would induce a self-reproducing change in what was present: for instance, two blocks would come together in such a way that specific pairs were formed. If we follow the unit of two blocks in a certain state, the unit itself reproduces. It reproduces itself out of the specific substrate it needs, just as all organisms do. The reason that it fails so ludicrously in being anything we would like to call living, even though it has the property of self-reproduction, is that it makes only one link. Organisms like ourselves contain many more than 10^{10} subtle and marvelously distinctive links. It is a quantitative question, which turns into something deeply qualitative as you watch. The other metaphors of life, such as flames which transfer from candle to candle, have some semblance to life. Far from being stupid, these metaphors are wonderfully insightful. But they do not go very far because you just get *flame* or *no-flame*, the flame is nearly the same no matter what the candle shape is to begin with, and so on. The amount of free energy involved in the few particular discriminations is tiny compared to the free-energy change of burning the whole candle. This suggests at once that living forms either must possess order on a scale small compared to ourselves, for example, cellular order, or they have order on some larger scale and, therefore, are vastly larger than ourselves, as is our whole society.

I cannot yet make this definition really hold tight, for there are problems concerning the uniqueness of the definition of the free energy which I don't want to discuss here. But I think it is a suggestive direction. I believe that the general characterization is right, and that there is a meaning to the evolution of life which is unique, objective, and determined though in a varied way by structure. Therefore, given the thermodynamic and chemical conditions, it is very hard to believe it will not with some probability occur. I cannot tell you when it first occurred. I can't tell you what range of conditions it can occupy. But there must be many ranges. All I'm saying is that if it is structural, then there must be domains pretty close to the one we deal with here, where not only it can occur, but it must occur. It has not been excluded in any thermodynamic fashion. That is my assertion. I would very much like to put down a number of the order of 1 for β.

What about evolution? It has been very nicely described in Ch. 6 in language quite different from the language we will use here. You have seen counterparts of the logical design of computational flow charts or of the computers themselves. The compilers which are realized in disk, tape, and electronic devices, with their nonlinear and combinatoric nature, are also present in biochemical equilibrium reactions with specific catalysts. The same graphic language, the same language of trees and branches seems to be able to describe both in a simple and in a very strong way. That is why we see these two examples of very complicated semiorganisms, the real living ones and the imitation computer-like ones as similar. Powerful combinatorics give rise to innumerable branched combinations which can therefore have very complicated states of behavior. The structure will be at the basis of a comparison between the two. The computers will suffer badly by my criterion—they will not be living things at all, because if you ask what free energy they use per bit of stored information, it is hopelessly high compared to kT. Now let us take 1 in.^{3} for a memory of 10^6 elements, or some value close to that. But this is not the case in a genetic macromolecule, or in the brain. There you almost get one bit for every few atomic bonds instead of one bit for every cubic micron. Designers can hardly do that well today in microcircuitry; nevertheless, $1 \, \mu^3$ holds 10^{12} bonds, so they have a long way to go. Since the complexity is an exponential function of this kind of combinatorics, there is really a gigantic gap between computers and flatworms or any other simple kind of organism. Computer experts have

Figure 10.2 A schematic presentation of the origins of living things. The first graph shows the population of structures—molecules, for instance—against some measure of size or complexity for a case where a free-energy flow has spread the composition to include species of high free energy. The second graph shows what happens when some autocatalytic process begins to increase the population beyond some level of complexity by further degrading the less elaborate material present. The last graph suggests how artifacts might be looked at in this process, a second and rarer, but more costly, bump.

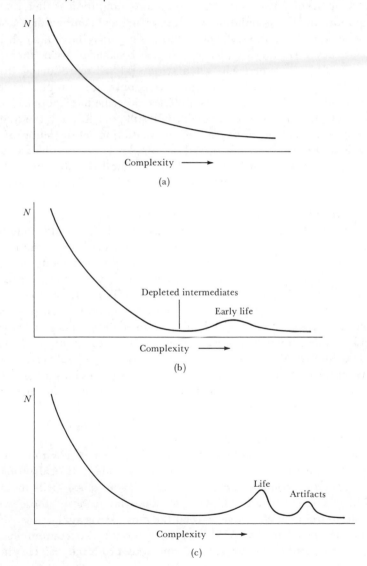

a long, long way to go. If they work hard, their machines might approach the intelligence of a human. But the human species is not one person, it is 10^{10} of them, and that is entirely a different thing. When they tell you about 10^{10} computers, then you can start to worry.

Division and Convergence

Now we shall take up another argument. Let us consider, as did Aronoff in Ch. 6, any thermodynamic system which is an open system, since energy is poured in from outside. It is a steady-state condition rather than a true equilibrium. The population will always include members of a molecular species that are quite difficult to form. They may have a small lifetime, but they form anyway; since they are broken up soon, there are not many of them. When you get a structural self-reproducing thing, an autocatalyst as he put it (though that is only one complex general class of models), what happens is that the intermediate population is depleted. The total free energy available will still pour through the system. But now there will tend to develop a differentiation by depletion of the intermediate states, a bimodal spread of complexity. If you take a cubic foot of sea water, you might very well find a small flounder in it. That is hopelessly far from the steady-state population. There is a great difference between that sample and plain old sea water. In the very old days, before there were any flounders or any organisms at all living in the sea, you would find no such big differences. The sea water may well have contained all sorts of molecules, such as purines and pyrimidines; such a population probably faded off with weight in a monotone and simple way. The present differentiation is one sign of the presence of complicated structures which contain their own information store. I suspect that the tendency of living forms is always to do this, to pass from heterotrophs to autotrophs, and gradually to increase the gap between life and nonlife. Now, of course, things are very different; in 1 mile^3 you could find a submarine full of crew members and software, a still more complex configuration. All the planets we are talking about have formed a population differentiation between the substrate and life. This will be one sign of the presence of this stored structure which I would like to call living forms. If that is the case, we can put a number like unity for the continual evolution probability.

We now come to the problem of n_{sp}, the number of communicative species. There, of course, we come to another serious statistical problem, which has caused controversy. Many expert evolutionists will disagree with what I say, for example, G. G. Simpson, who authored those works on evolution from which I have learned the most. He is against the whole proposition that evolution could ever have produced a communicative species in any other way, except the one that produced us! He shows that certain mud fish turned to the left not to the right at certain times and thus walked out a little farther on dry land. The fish and his posterity had first this experience, that experience, and the other experience, pretty soon, here we humans are. Of course, if you look at it this way then the argument is absolutely irrefutable; on the other hand it is a rather curious way to look at it. If you ask yourself how you came to be in a certain place and what the probability is that you could do it again exactly in the same way, I venture to say you could not manage it. By exactly the same way, I mean a space-time diagram which would fit with precision, one trip on top of the other. That means you have to hit the same traffic lights, the same curves, the same turns. It could not happen. The probability of any single given path is small, but there are many paths to the same approximate position. That is the only question open: Are there or are there not many evolutionary paths to the same approximate human ecological position? Of course we do not surely know, we do not have a sufficiently strong theory to answer that. But there is a very strong analogical argument of other successful ecological positions.

Let us examine the famous example of those creatures who make a living by swimming in the sea at 20 or 30 mph, catching other beasts that are smaller than 30 cm, swift predators on small coastal fish. I know of three species which have made a living in just that way in the course of earth history. Two species which are still extant are the tuna, a fish, and the dolphin, a mammal; the third species is the ichthyosaur, a reptile. These animals have very little to do with each other biochemically and phylogenetically. Their histories are quite distant, separated by 100 million yr for each step. Yet if you look at the external photograph of the three forms, they all look about the same. They look rather like torpedos; in fact, if we imagined a self-reproducing, self-nourishing manmade torpedo, it would be a steel and aluminum object of the same form, though it is still not yet developed. Such convergence is an example of there possibly being many evolutionary paths to the same approximate ecological position. We do not need to enlarge on the Australian example of the pseudomammals, the marsupials, or that of the Galapagos birds, which evolved out of one finch species into a whole number of bird species adapted to different niches.

We will argue that there is one way of making a living which is extremely successful ecologically (at least for a time). It is our own human way by which we spread over all of the latitudes and longitudes of the earth; our biomass is greater than that of any but a very few other species. We must be regarded biologically as very successful. Beginning somewhere in the East African rift valley as very modest forest primates, we eventually found this way of living. It is probably going to be reached more than once by different paths: the only difference is time. The whole argument of separating population peaks works at this level, too, because if we see a species now tending to become competitive with us, we will not allow them to remain in that position; we will either dominate them, or have them join us. We will not allow them to remain in an uncontrolled but competitive position. That is quite plain and is a theorem well known in ecology. I feel that indeed our way of living lies on a plateau in species phase space, not just at a single point. Many paths lead to it, approximately. So I do not hesitate to put down the order of 1 for that number n_{sp}. (I do not put down 1 million because we have not seen more than one, or three, maybe, in the history of our earth and that is a good guide.) But I think it is of the order of 1; I would like to put it that way. If you make it $1/10$, I would not argue that to be wrong.

Let me make one additional remark on this subject. This convergence, as it is called in evolution, is not I admit, any proof that evolution could have produced communicative species by other paths, but it is very suggestive indeed. It is an interesting question to ask yourself. I would like to see a serious investigation made of it (it has been done very imaginatively, again by science fiction writers!) by a biologist who knows the situation. Suppose that there had not been any primates. What existing order of mammalia—primates absent—is most likely to develop in time, in a few $\times\ 10^8$ yr, into a manipulative, environment-changing form? The raccoon strain is now very close to being what those early lemurian primates were more than 50 million years ago when the order of primates began to be differentiated in that way. We can hardly look at an omniverous, intelligent raccoon with the manipulative behavior of his front paws without seeing that there is implied in his being, provided the right sort of things happened, the possibility that his kin could find their way to the plateau of technology. Perhaps some day we will know.

The Feedback of Consciousness

The last number is T_c/T. Although it is the most difficult question, we shall not spend a lot of time discussing it. It cannot be smaller than 10^{-8} or 10^{-9}; that is to say, if we last communicatively only 50 or 100 yr, while the evolutionary time is $5\ \times\ 10^9$ yr, we define a minimal value. Then we are alone; we are a transitory phenomenon on the galactic scene; complex radio signals, quite apart from all other natural processes, are going to disappear forever very soon, because we will stop sending within 50 to 100 yr. That is a most pessimistic point of view; it leads to the unique answer that we are now alone in the galaxy. It could certainly be that way. On the other hand, it could also be that Darwin's evolution stops with environment manipulation or biological evolution, and that only social evolution continues until it reaches a steady state. The other extreme possibility is that we continue forever to be communicative. In this case, the duration T_c is more likely governed by prodigious changes in the natural environment, probably more severe than the ice ages. That probably means the lifetime of stars. So we use 10^9, or something like 10^9, yr for T_c, and the fraction grows very large indeed. If we take that large number, then within 1,000 light-years or so, it is very likely that we should find one such ancient, communicative species. The only question then is how to locate it.

I will close with a few remarks about this question, which is the central point. We must remember that we ourselves have been here, let us say, one million years, or we can say 100 yr. (The difference is within our error!) If we chose 100 yr, we are talking about radio transmission, while 1 million yr is the proper number if we are to talk about the species. But compare this with 5×10^9 yr, the evolutionary time of the whole environment, star, earth, and the diverse species of life. Either time is but an instant of geologic time. If we plot the earth's population of radio transmitters as a function of time, against the history of the earth, we have a perfect step function. Zero radio transmitters existed until the instant *now*, then the curve rises up to however many there are now, and sooner or later the curve will plateau off. So it is effectively the true step function. Of course, the rise of that step function contains all of human evolution and all of human history; from our provincial point of view, it is a very elaborate step, but from the point of view of probability theory, it is a step function like any other. What we say is only that this step is not synchronized in any way with cosmic beginnings. It took 5 billion yr, but it might equally have taken 4 billion, or it might have taken 6 billion; this is clear from the chancey nature of the whole process which involves millions of chance acts. Therefore it seems quite likely that there is no synchrony anywhere; the stars do not start at the same time, and it is probable that civilizations are spread uniformly around the time which corresponds to the time the stars last, billions of years, with a knife-edge rise of millions or hundreds of years, whichever value you use. However you take it, you are going to find that our knife edge is not synchronized to any other. If it is not synchronized, then it is either behind or ahead. Either we are ahead of everybody so that we are unique, or we are behind many; if we are behind, it is very unlikely that we are just on the edge, so that there are two independent knife edges coming together! It is far more likely that we are well behind, as in other cases, well ahead, but *well behind* means a time that is *long compared to the whole duration of human culture.* That means an enormous development; nobody is like us. There is no point in the commonplace science fiction stories that tells you they have city-state leagues, royal alliances, and weapons very similar to ours, the same game played out with different names. As a point of difference, perhaps they have furry ears! That view is not going to work. They are going to be totally different in culture. Either they have completely transcended our technology, or they are still in a lemurian stage. Everything else is extremely improbable.

That is a hard argument to beat because it depends on one large number—the ratio of 1 billion yr either to 100 yr or to millions of years: 10^7 or 10^4. Both ratios are very large compared to 1. That is the point that tells. We must concede that if there are other beings, they are very much farther along in technology than we are. That is not to say that they are infinitely powerful, rather, they have reached the limitations of technology. We do not know what those limits are, but they must exist. Those fellows are at the limits. Those limitations presumably do not imply a mere factor of ten here and there in the kind of things we do. We might as well assume that they have much better radio transmitters than we have, so we should let them do the hardest part of the work, which is the rationale for the notion that we should listen instead of transmit. We listen in the microwave region; we try to find the targets. There are tens of millions of targets! Dr. Oliver gave a very detailed account of the procedures by which you might build up a search system. Someday we are going to try something like that, provided we can agree on the general sort of acquisition signals expected.

Finally, both Von Hoerner and Hoyle have argued, I think quite persuasively, that there is probably a great feedback system in this communications circuitry. Namely, if we build any link, it is extremely likely that this action increases the time that the link endures. If any other beings exist and have formed their links, this increases the chance that we can too. The system is tightly fed back; either there are no links at all or there are many, in a very nonlinear way. The chance of success enhances the chance of trying, and further success enhances it still more. If there are many possibilities, they have mostly been discovered; we are simply the latest ones to come in on this party line.

It will be many or none. Let us hope for many, for contact ahead in a time comparable to the duration of our readiness, mere decades since Jansky, Reber, and Hey.

A Bibliography on Interstellar Communication

Linda D. Caren[1, 2], Eugene F. Mallove[3], and Robert L. Forward[3]

The bibliographic entries have been sorted into five major categories with their subdivisions, then listed alphabetically by author within each subdivision. Each source appears in only one subdivision in the bibliography. Of course, some sources have fit less easily than others into a single category, but an effort has been made to match the predominant content of an entry with a given subject category. The subdivisions of the bibliography are as follows:

I. PROBABILITY OF EXTRASOLAR INTELLIGENCE
   A. Probability of Extrasolar Planets
   B. Detection of Extrasolar Planets
   C. Life-Supporting Regions
   D. The Origin and Evolution of Extrasolar Life

II. METHODS OF COMMUNICATING WITH EXTRASOLAR INTELLIGENCE
   A. Searching for Signals from Extrasolar Intelligence
   B. Interstellar Communication Languages

III. PHILOSOPHICAL, PSYCHOLOGICAL, AND SOCIOLOGICAL QUESTIONS OF INTERSTELLAR COMMUNICATION AND EXTRASOLAR INTELLIGENCE

[1] I wish to thank Dr. Keith Kvenvolden, Chief, Chemical Evolutionary Branch, Planetary Biology Division, Ames Research Center, NASA, and the Ames Research Center, NASA, for greatly facilitating the compilation of this bibliography.
[2] Planetary Biology Division, Ames Research Center, NASA, Moffett Field, California 94035.
[3] Exploratory Studies Department, Hughes Research Laboratories, Malibu, California 90265.

IV.   BIBLIOGRAPHIES, COMPENDIA, AND BOOKS COVERING
      MULTIPLE TOPICS OF EXTRASOLAR INTELLIGENCE
      AND INTERSTELLAR COMMUNICATION
      A.   Bibliographies
      B.   Compendia
      C.   Multiple topic books
V.   MISCELLANEOUS

Entries on the origin and evolution of possible extrasolar life were selected from the very large body of literature which exists on extraterrestrial
biology. Our general criterion for selection was to exclude detailed biochemical discussions but to include works which attempt to assess the
probability of the emergence of intelligent life from presupposed primitive
chemical evolution. Similarly, we have not included the large number
of references on the origin of organic matter in meteorites, possibility
of life in our solar system, techniques and hardware which could be used
to detect such life, and possible methods of interstellar transport. We
hope that this gathering of information will be of use in generating new
ideas for methods to bridge the interstellar gulf.

I.   Probability of Extrasolar Intelligence

A.   *Probability of Extrasolar Planets*
Alfvén, Hans: Non-Solar Planets and the Origin of the Solar System,
    Nature, vol. 152, no. 3868, p. 721, 1943.
Banerji, A. C.: Non-Solar Planetary Systems, *Nature,* vol. 153, no. 3895,
    p. 779, 1944.
Brown, Harrison: Planetary Systems Associated with Main-Sequence
    Stars, *Science,* vol. 145, no. 3637, pp. 1177–1181, 1964.
The CETI (Communication with Extraterrestrial Intelligence) Resolu-
    tions, *Astronautics and Aeronautics,* vol. 9, no. 11, pp. 35, 57, 1971.
Dark Companions, *Time,* vol. 42, no. 5, pp. 46–47, 1943.
Dole, S. H.: Computer Simulation of the Formation of Planetary Systems,
    Icarus, vol. 13, pp. 494–508, 1970.
Ewing, Ann: Life in All Solar Systems?, *Sci. News Lett.,* vol. 86, p. 199,
    1964.
Huang, Su-Shu: Occurrence of Planetary Systems in the Universe as a
    Problem in Stellar Astronomy, in A. Beer (ed.), "Vistas in Astronomy,"
    vol. 11, pp. 217–263, Pergamon Press, New York, 1969.
Hunter, A.: Non-Solar Planets, *Nature,* vol. 152, no. 3846, pp. 66–67,
    1943.
Jeans, H. H.: Non-Solar Planetary Systems, *Nature,* vol. 152, no. 3868,
    p. 721, 1943.

Kumar, S. S: On Planets and Black Dwarfs, *Icarus,* vol. 6, pp. 136–137, 1967.

———: Planetary Systems, in "The Emerging Universe: Essays on Contemporary Astronomy," pp. 25–34, University Press of Virginia, Charlottesville, 1972.

Lawton, A. T.: The Nearest Other Solar System?, *Spaceflight,* vol. 12, no. 4, pp. 170–173, 1970.

Ley, Willy: The Planets of Other Stars, in "Missiles, Moonprobes, and Megaparsecs," chap. 19, pp. 165–175, New American Library, Inc., New York, 1964.

Life on a Billion Planets?, *Time,* vol. 71, pp. 42–43, March 3, 1958.

Macvey, John W.: Planets of Other Stars, *Planetarium,* vol. 1, pp. 116–118, 1968.

Other Beings on Other Planets?, *Life,* vol. 44, no. 1, p. 6, 1958.

Ovenden, Michael W.: Extrasolar Planets, in "Recent Development in Astronomy," *J. Brit. Interplanetary Soc.,* vol. 7, no. 2, pp. 78–80, 1948.

Proell, Wayne: The Probability of Planets in Solar Space, *The J. of Spaceflight,* vol. 5, no. 9, pp. 1–7, Chicago Rocket Society, 1953.

Russell, Henry Norris: Anthropocentrism's Demise, *Sci. Amer.,* vol. 169, no. 1, pp. 18–19, 1943.

Sen, H. K.: Non-Solar Planetary Systems, *Nature,* vol. 152, no. 3864, pp. 601–602, 1943.

Serviss, G. P.: Are There Planets among the Stars?, *Popular Sci.,* vol. 52, pp. 171–177, Dec. 1897.

Young, Andrew T.: The Occurrence of Planets, *Science,* vol. 148, no. 3669, p. 532, 1965.

B. *Detection of Extrasolar Planets*

Berger, William J.: Celestial Iconospherics, the Ultimate Astronomy, *Jet Propulsion,* vol. 28, no. 5, pp. 337–338, 1958.

Bracewell, Ronald N.: Interstellar Probes, in Cyril Ponnamperuma and A. G. W. Cameron (eds.), "Interstellar Communication: Scientific Perspectives," chap. 7, Houghton Mifflin Company, Boston, 1974.

Bramson, Herbert J.: "The Detection of Interstellar Planets: A Feasibility Study," Hughes Aerospace Group TM-825, 1965.

Briggs, M.H.: The Detection of Planets at Interstellar Distances, *J. Brit. Interplanetary Soc.,* vol. 17, pp. 59–60, 1959.

Graf, E. R., F. A. Ford, M. D. Fahey, and R. J. Coleman: A Proposed Experimental Method for Determining the Existence of Other Planetary Systems, in Harry Zuckerberg (ed.), *Adv. in the Astronautical Sciences,* vol. 24 ("Exploitation of Space for Experimental Research"), pp. 89–94, Proc. 14th Ann. American Astronautical Soc. Meeting, May 13–15, Dedham, Mass., 1968.

van de Kamp, Peter: Stars or Planets?, *Sky and Telescope,* Dec. 1944; also in Thornton Page and Lou Williams Page (eds.), "The Origin of the Solar System," pp. 176–180, The Macmillan Company, New York, 1966.

———: Planetary Companions of Stars, in A. Beer (ed.), "Vistas in Astronomy," vol. 2, pp. 1040–1048, Pergamon Press, New York, 1956.

———: Barnard's Star As an Astrometric Binary, *Sky and Telescope,* vol. 26, no. 1, p. 8, 1963.

———: Astrometric Study of Barnard's Star from Plates Taken with the 24-inch Sproul Refractor, *Astronomical J.,* vol. 68, no. 7, pp. 515–521, 1963.

———: The Discovery of Planetary Companions of Stars, *Yale Sci. Magazine,* vol. 38, no. 3, pp. 6–8, 1963.

———: The Search for Perturbations in Stellar Proper Motions, in A. Beer and K. Strand (eds.), "Vistas in Astronomy," vol. 8, pp. 215–218, Pergamon Press, New York, 1966.

———: Parallax, Proper Motion, Acceleration, and Orbital Motion of Barnard's Star, *Astronomical J.,* vol. 74, no. 2, pp. 238–240, 1969.

———: Alternate Dynamical Analysis of Barnard's Star, *Astronomical J.,* vol. 74, no. 6, pp. 757–759, 1969.

———: Extrasolar Planetary Systems, *Adv. in Space Sci. and Tech.,* vol. 11, pp. 437–441, 1972.

Lawton, A. T.: Photometric Observation of Planets at Interstellar Distances, *Spaceflight,* vol. 12, no. 9, pp. 365–373, 1970.

Leinwoll, Stanley: Which Stars Have Planets?, *Analog,* vol. 71, no. 2, pp. 9–16, 91–93, 1963.

Notes and News: The Detection of Extra-Solar Planets, report on Otto Struve proposal, *J. Brit. Interplanetary Soc.,* vol. 12, no. 1, pp. 77–78, 1953.

Roman, Nancy Grace: Planets of Other Suns, *Astronomical J.,* vol. 64, no. 1273, pp. 344–345, 1959.

Rosenblatt, Frank: A Two-Color Photometric Method for Detection of Extra-Solar Planetary Systems, *Icarus,* vol. 14, no. 1, pp. 71–93, 1971.

Spitzer, Lyman, Jr.: The Beginnings and Future of Space Astronomy, *Amer. Sci.,* vol. 50, pp. 473–484, 1962.

Sullivan, Walter: Message of the Pulsars, *New York Times,* May 26, 1968, sect. 4, p. 12.

Two Planet Solar System, *Sci. Amer.,* vol. 220, no. 6, p. 58, 1969.

C.   *Life-supporting Regions*

Anderson, Poul: Artificial Biosphere, correspondence in *Science,* vol. 132, pp. 251–252, 1960.

Bún, Thomas P. and Flávio A. Pereira: 'Biospheric Index,' a Contribution to the Problem of Determination of the Existence of Extra-Solar Planetary Biospheres, *Proc. 8th Int. Astronautical Congress,* Barcelona, 1957, Springer-Verlag OHG, Vienna, 1958, p. 63.

Cameron, A. G. W.: Stellar Life Zones, in A. G. W. Cameron (ed.), "Interstellar Communication," pp. 107–110, W. A. Benjamin, Inc., New York, 1963.

———: Planetary Systems in the Galaxy, in Cyril Ponnamperuma and A. G. W. Cameron (eds.), "Interstellar Communication: Scientific Perspectives," Houghton Mifflin Company, Boston, 1974.

Campbell, John W.: Where Did Everybody Go?, in "Collected Editorials from Analog," selected by Harry Harrison, pp. 232–243, Doubleday and Company, Inc., Garden City, N.Y., 1966.

Condon, Edward U.: Intelligent Life Elsewhere, "Scientific Study of Unidentified Flying Objects," pp. 26–33, *New York Times,* 1969.

Dole, Stephen H.: "Habitable Planets for Man," American Elsevier Publishing Company, Inc., New York, 1970.

——— and I. Asimov: "Planets for Man," Random House, Inc., New York, 1964.

Donahoe, F. G.: On the Abundance of Earth-Like Planets, *Icarus,* vol. 5, no. 3, pp. 303–304, 1966.

Dyson, Freeman, J.: Artificial Biosphere, correspondence in *Science,* vol. 132, pp. 252–253, 1960.

Ecosphere May Shape Life on Distant Planets, *Sci. News Lett.,* vol. 80, no. 17, p. 272, 1961.

Ehricke, Krafft A.: Astrogenic Environments — The Effect of Stellar Spectral Classes on the Evolutionary Pace of Life, *Spaceflight,* vol. 14, no. 1, pp. 2–14, 1972.

Ewing, Ann: Tiny 'Stars' with Life, *Sci. News Lett.,* vol. 81, no. 21, p. 323, 1962.

———: Life on Tiny, Dark Stars, *Sci. News Lett.,* vol. 82, no. 3, p. 39, 1962.

Fesenkov, V. G.: Conditions of Life in the Universe, *Priroda,* no. 1.70, pp. 20–27 (in Russian), 1970.

———: Life Conditions in the Universe, in A. A. Imshenetskii (ed.), "Extraterrestrial Life and Its Detection Methods," pp. 7–16 (in Russian), 1970.

———: Conditions for Life in the Universe, "Soviet Studies of the Universe," pp. 1–16, Joint Publications Research Service, Washington, D.C., 1970.

Gadomski, Jan: Die Sternenökospharen im Radius von 17 Lichtjahren um die Sonne, *Proc. 8th Int. Astronautical Congress,* Barcelona, 1957, pp. 127–136 (in German), Springer-Verlag OHG, Vienna, 1958; The Stellar Ecospheres within a Radius of 17 Light Years around the Sun, translation No. T115, Rand Corporation, Santa Monica, Calif., June 11, 1959.

———: Fünf Arten von ökosphärischen Planeten, (Five Types of Ecospheric Planets), *Proc. 9th Int. Astronautical Congress,* Amsterdam, 1958, pp. 785–793 (in German), Springer-Verlag OHG, Vienna, 1959.

———: Die Okosphären der veränderlichen Sterne, (The Ecospheres of Pulsating Stars), *Proc. 11th Int. Astronautical Congress,* Stockholm, 1960, pp. 108–113 (in German), Springer-Verlag OHG, Vienna, 1961.

———: Ecospheric Consequences of the Haselgrove-Hoyle-Schwarzschild Theory Concerning the Evolution of the Sun, *Proc. 13th Int. Astronautical Congress,* Varna, 1962, pp. 54–58, Springer-Verlag OHG, Vienna, 1964.

Huang, Su-Shu: Life Supporting Regions in the Vicinity of Binary Systems, in A. G. W. Cameron (ed.), "Interstellar Communication," pp. 93–101, W. A. Benjamin, Inc., New York, 1963; *Astronomical Soc. of the Pacific, Publications,* vol. 72, pp. 106–114, 1960.

———: The Sizes of Habitable Planets, in A. G. W. Cameron (ed.), "Interstellar Communication," pp. 102–106, W. A. Benjamin, Inc., New York, 1963; *Astronomical Soc. of the Pacific, Publications,* vol. 72, pp. 489–493, 1960; NASA TN–D–499.

Leiber, Fritz: Homes for Men in the Stars, *Sci. Digest,* vol. 58, pp. 53–57, Sept. 1965.

Ley, Willy: "Beyond the Solar System," The Viking Press, Inc., New York, 1964.

Life On Other Planets, *New York Times,* July 12, 1959, sect. 4, p. 9.

Little Inhabited Stars, *Time,* vol. 73, p. 65, April 27, 1959.

Litynski, Zygmunt: Life on Other Planets: Views of a Polish Scientist, *Sci. Digest,* vol. 50, pp. 71–75, Aug. 1961.

Maddox, John: Artificial Biosphere, correspondence in *Science,* vol. 132, pp. 250–251, 1960.

Maude, A. D.: Life in the Sun, in I. J. Good (ed.), "The Scientist Speculates," pp. 240–247, William Heinemann, Ltd., London, 1962.

Moore, Patrick and Francis Jackson: Planets of Other Stars, "Life in the Universe," pp. 106–114, W. W. Norton and Company, Inc., New York, 1962.

Öpik, E. J.: Stellar Planets and Little Dark Stars As Possible Seats of Life, *Irish Astronomical J.,* vol. 6, p. 290, 1964.

The Prevalence of Planets and the Probability of Life, *Time,* vol. 84, no. 13, pp. 49–50, 1964.

del Ray, Lester: Let There be Life, "The Mysterious Sky," chap. 18, pp. 171–177, Chilton Books Company, Philadelphia, 1964.

Sagan, Carl: An Introduction to the Problem of Interstellar Communication, in Cyril Ponnamperuma and A. G. W. Cameron (eds.), "Interstellar Communication: Scientific Perspectives," Houghton Mifflin Company, Boston, 1974.

Shapley, Harlow: Concerning Life on Stellar Surfaces, in I. J. Good (ed.), "The Scientist Speculates," pp. 225–233, William Heinemann, Ltd., London, 1962.

Sloane, Eugene: Artificial Biosphere, correspondence in *Science,* vol. 132, p. 252, July 1960.

Strughold, H.: The Ecosphere in the Solar Planetary System, in *Proc. 7th Int. Astronautical Congress,* Sept. 17–22, 1956, Rome, pp. 277–288.

D. *The Origin and Evolution of Extrasolar Life*

Agnew, I.: Life on Other Planets?, *Sci. Digest,* vol. 72, p. 86, Sept. 1972.

Amidei, R.: Ingredients for Life in Outer Space, *Sci. Digest,* vol. 70, pp. 24–29, Sept. 1971.

Animal, Vegetable, or . . . , *Newsweek,* vol. 56, p. 78, Nov. 28, 1960.

Arbib, Michael: The Likelihood of the Evolution of Communicating Intelligences on Other Planets, in Cyril Ponnamperuma and A. G. W. Cameron (eds.), "Interstellar Communication: Scientific Perspectives," Houghton Mifflin Company, Boston, 1974.

Are There Spores in Space?, *Sci. Digest,* vol. 48, inside back cover, Aug. 1960.

Arrhenius, Svante: The Transmission of Life from Star to Star, *Sci. Amer.,* vol. 96, p. 196, March 2, 1907.

———: The Spreading of Life through the Universe, "Worlds in the Making," chap. 8, pp. 212–230, Harper and Brothers, New York, 1908.

Asimov, Isaac: A Science in Search of a Subject; We, the In-Betweens; Anatomy of a Martian, "Is Anyone There?," chaps. 20, 21, and 23, pp. 183–196, 207–211, Doubleday and Company, Garden City, N. Y., 1967.

Austin, R. R.: Extraterrestrial Life, correspondence in *Spaceflight,* vol. 4, no. 5, p. 176, 1962.

Baidins, A.: correspondence in *Analog,* vol. 86, no. 3, p. 173, 1970.

Baur, Franz: Cosmic Aspects of Organic Evolution — A Supplement to the Contribution by Prof. G. Simpson, *Naturwissenschaftliche Rundschau,* vol. 22, pp. 167–168 (in German), 1969.

Berget, A.: Appearance of Life on Worlds and the Hypothesis of Arrhenius, *Smithsonian Inst. Ann. Rep.,* pp. 543–551, 1912–1913.

Bernstein, Jeremy: Life in the Universe, *New Yorker,* pp. 117–120, 123–126, 129–132, 135–138, 141–142, May 28, 1966.

Berrill, N. J.: The Search for Life, *The Atlantic,* vol. 212, no. 2, pp. 35–40, 1963.

Bieri, Robert: Humanoids on Other Planets, *Amer. Sci.,* vol. 52, pp. 452–458, 1964.

Biological Problems of Space Flight, a report on Prof. J. B. S. Haldane's lecture, *J. Brit. Interplanetary Soc.,* vol. 10, no. 4, pp. 154–158, 1951.

Bova, Ben: Life Cycles, *Analog,* vol. 89, no. 4, pp. 5–6, 174–175, 1972.

Bracewell, R. N.: Life in the Galaxy, in A. G. W. Cameron (ed.), "Interstellar Communication," pp. 232–242, W. A. Benjamin, Inc., New York,

1963; in S. T. Butler and H. Messel (eds.), "A Journey through Space and the Atom," Nuclear Research Foundation, Sydney, Australia, 1962.

Briggs, M. H.: Terrestrial and Extraterrestrial Life, *Spaceflight,* vol. 2, pp. 120–121, 1959.

———: Other Astronomers in the Universe?, *Southern Stars* (New Zealand), vol. 18, pp. 147–151, 1960.

Buhl, David and Cyril Ponnamperuma: Interstellar Molecules and the Origin of Life, *Space Life Sci.,* vol. 3, no. 2, pp. 157–164, 1971.

Cade, C. M.: Are We Alone in Space?, *Discovery,* vol. 24, pp. 27–29, 32–34, Apr. 1963.

Calvin, Melvin: Communication: From Molecules to Mars, *A. I. B. S. Bull.,* pp. 29–44, Oct. 1962.

Campbell, John W.: The Nature of Intelligent Aliens, *Analog,* vol. 80, no. 2, pp. 5–7, 172–178, 1967.

Chaikin, G. L.: A Transitional Hypothesis Concerning Life on Interstellar Bodies, *Popular Astronomy,* vol. 59, no. 1, pp. 50–51, 1951.

Chance of Life, *Chem. and Eng. News,* vol. 49, pp. 8–9, March 29, 1971.

Cohen, D.: Life on Earth and Elsewhere, *Sci. Digest,* vol. 59, pp. 92–93, March 1966.

Cook, Rick: Life as We Don't Know It, *Analog,* vol. 86, no. 3, pp. 39–59, 1970.

Dean, J. C.: The Transmission of Life from Star to Star, *Sci. Amer.,* vol. 96, p. 331, April 20, 1907.

Diamond, Edwin: Is There Life in Outer Space?, *Newsweek,* Feb. 22, 1960.

Did Life Reach Earth on a Comet's Bright Trail?, *Newsweek,* vol. 6, p. 39, Dec. 14, 1935.

Did the Seeds of Life Come from Space?, *Sci. News,* vol. 101, p. 231, April 8, 1972.

Dobzhansky, Theodosius: Darwinian Evolution and the Problem of Extraterrestrial Life, *Perspectives in Biol. Med.,* vol. 15, no. 2, pp. 157–175, 1972.

Do Other 'Humans' Live?, *Newsweek,* vol. 52, p. 56, Nov. 17, 1958.

Earth Life Not from Space, *Sci. News Lett.,* vol. 81, p. 4, Jan. 6, 1962.

Eiseley, Loren C.: Is Man Alone in Space?, *Sci. Amer.,* vol. 189, pp. 80–82, 84, 86, July 1953.

Ensanian, Minas: Does Life Exist Elsewhere in the Universe?, *Proc. of the 5th Space Congress,* Cocoa Beach, Fla., March 11–14, 1968, 25.5-1–25.5-20.

Evans, Selby: A New Thought about Life on Other Planets, *Amer. Mercury,* vol. 86, no. 413, pp. 83–88, 1958.

The Evolution of Life in the Universe, a report on a lecture by Prof. J. D. Bernal, *J. Brit. Interplanetary Soc.,* vol. 12, no. 3, pp. 114–118, 1953.

Extraterrestrial Life, *Encyclopaedia Britannica,* vol. 13, pp. 1083M–1088, William Benton, Publisher, Chicago, 1972.

Extraterrestrial Neighbors? *Chemistry,* vol. 46, p. 16, Feb. 1973.

Fesenkov, V., J. Shklovskiy, A. Pasynskiy, and B. Lyapunov: Civilization on Other Planets?, *USSR,* vol. 1, no. 88, pp. 46–47, 1964.

Finney, John W.: Biologist Backs Space Plan Foes, *New York Times,* June 9, 1963, p. 21.

Firsoff, V. A.: An Ammonia-Based Life, *Discovery,* vol. 23, no. 1, pp. 36–42, 1962.

———: Possible Alternative Chemistries of Life, *Spaceflight,* vol. 7, no. 4, pp. 132–136, 1965.

Fox, Sidney W.: Humanoids and Proteinoids, *Science,* vol. 144, no. 3621, p. 954, 1964.

Gardner, Martin: The Origin of Life, "The Ambidextrous Universe," chap. 15, pp. 129–139, New American Library, Inc., New York, 1969.

———: Plants and Animals, "The Ambidextrous Universe," chap. 7, pp. 55–65, New American Library, Inc., New York, 1969.

Gazenko, O. G.: Life in Outer Space, *Priroda,* no. 10.70, pp. 80–82 (in Russian), 1970.

Goldmine in the Sky, *Newsweek,* vol. 77, p. 56, April 26, 1971.

Good, I. J.: Interstellar Communication for Chemical Research, in I. J. Good (ed.), "The Scientist Speculates," pp. 239–240, William Heinemann, Ltd., London, 1962.

———: The Human Preserve, *Spaceflight,* vol. 7, no. 5, pp. 167–170, 1965.

Heuer, K.: Men of Other Planets, *Sci. Digest,* vol. 30, pp. 9–13, July, 1951.

Horne, R. A.: On the Unlikelihood of Non-Aqueous Biosystems, *Space Life Sci.,* vol. 3, no. 1, pp. 34–41, 1971.

Horowitz, Norman H.: The Biological Significance of the Search for Extraterrestrial Life, in J. S. Hanrahan (ed.), "The Search for Extraterrestrial Life," *Adv. in the Astronautical Sciences,* vol. 22, pp. 3–13, American Astronautical Society, 1967.

Howells, William: Would Other 'Humans' Look Like Us?, *Sci. Digest,* vol. 47, no. 1, pp. 53–58, 1960.

———: The Evolution of 'Humans' on Other Planets, *Discovery,* vol. 22, no. 6, pp. 237–241, 1961.

Huang, Su-Shu: Occurrence of Life in the Universe, in A. G. W. Cameron (ed.), "Interstellar Communication," pp. 82–88, W. A. Benjamin, Inc., New York, 1963; *Amer. Scientist,* vol. 47, pp. 397–402, 1959.

———: The Problem of Life in the Universe and the Mode of Star Formation, in A. G. W. Cameron (ed.), "Interstellar Communication," pp. 89–92, W. A. Benjamin, Inc., New York, 1963; *Astronomical Soc. of the Pacific, Publications,* vol. 71, pp. 421–424, 1959.

———: Life Outside the Solar System, *Sci. Amer.,* vol. 202, no. 4, pp. 55–63, 1960.

———: Some Astronomical Aspects of Life in the Universe, *Sky and Telescope,* vol. 21, no. 6, pp. 312–316, 1961.

——— and R. H. Wilson: Astronomical Aspects of the Emergence of Intelligence, IAS Paper No. 63–48, presented at the 31st Ann. Meeting of the Institute of the Aerospace Sciences, New York City, Jan. 21–23, 1963.

Jastrow, Robert: A Message, "Red Giants and White Dwarfs: The Evolution of Stars, Planets, and Life," chap. 12, pp. 138–140, Harper and Row, Publishers, Inc., New York, 1967.

Jeans, James: Is There Life on the Other Worlds?, *Science,* vol. 95, no. 2476, pp. 589–592, 1942.

Jones, Sir Harold Spencer: Beyond the Solar System, "Life on Other Worlds," Mentor Books, New American Library, Inc., New York, 1956.

Kahn, F. D.: Life in the Universe, "The Emerging Universe: Essays on Contemporary Astronomy," University Press of Virginia, Charlottesville, pp. 71–89, 1972.

Kavaler, Lucy: Can Life Begin on a Cold Planet?, "Freezing Point: Cold as a Matter of Life and Death," chap. 20, pp. 345–362, The John Day Company, Inc., New York, 1970.

Keilin, D.: Anabiosis and Its Bearing on Problems of the Origin and the Continuity of Life on the Earth, *Proc. Royal Soc. London,* ser. B, vol. 150, pp. 180–181, 1959. In "The Problem of Anabiosis or Latent Life: History and Current Concept," *Proc. Royal Soc. London,* ser. B, vol. 150, pp. 149–191, 1959.

Keller, Eugenia: Origin of Life: Part III. On Other Planets?, *Chemistry,* vol. 42, no. 4, pp. 8–13, 1969.

Kleczek, Josip: Life on the Planets of Other Suns, "To the Near and Distant Universe," translation from Czech, pp. 411–427, Air Force Systems Command, Wright-Patterson AFB, Ohio, Foreign Technology Division, FTD-TT-62/7–1 + 2 + 3 + 4, AD-413009, July 16, 1963.

Kopal, Zdenek: Life in the Universe?, "Man and His Universe," William Morrow and Company, Inc., New York, pp. 300–308, 1972.

Kreifeldt, J. G.: A Formulation for the Number of Communicative Civilizations in the Galaxy, *Icarus,* vol. 14, no. 3, pp. 419–430, 1971.

Lafleur, L. J.: Surface Gravity and Behavior, *Popular Astronomy,* vol. 51, pp. 197–202, 1943.

Langer, R. M. and T. E. Stimson: Is There Life Among the Stars? *Popular Mechanics,* vol. 78, pp. 82–85, 1942.

Larson, Carl A.: Strong Poison 2, *Analog,* vol. 89, no. 4, pp. 73–84, 1972.

Laurence, William L.: Experts Hold Probability of Life Elsewhere in Universe is Great, *New York Times,* sect. 4, p. 9, March 5, 1961.

———: Beings Elsewhere—The Creatures on Other Planets Might Look Like Centaurs, *New York Times,* sect. 4, p. 8, July 9, 1961.

———: Tests Show Primitive Atmosphere Can Create Living Systems, *New York Times,* sect. 4, p. 11, Sept. 9, 1962.

Lederberg, Joshua: Exobiology: Approaches to Life Beyond the Earth, *Science,* vol. 132, pp. 393–400, 1961.

Ley, Willy: What Will Space People Look Like?, *Sci. Digest,* vol. 43, pp. 61–64, Feb. 1958.

———: Life in the Universe and the Concept of the Ecosphere, "Watchers of the Skies," chap. 22, pp. 478–500, The Viking Press, Inc., New York, 1963.

———: Let's Build an Extraterrestrial, "Another Look at Atlantis and Fifteen Other Essays," pp. 140–154, Doubleday and Company, Inc., Garden City, N.Y., 1969.

Libby, W. F.: Life in Space, *Space Life Sciences,* vol. 1, pp. 5–9, 1968.

Life Beyond the Earth?, *Sci. Digest,* vol. 69, p. 51, March 1971.

Life Elsewhere?, *New York Times,* sect. 4, p. 7, Oct. 15, 1961.

Life Out There, *Newsweek,* vol. 76, p. 118, Dec. 14, 1970.

Life's Spread Through the Universe, in Thornton Page and Lou Williams Page (eds.), "The Origin of the Solar System," pp. 308–310, The Macmillan Company, New York, 1966; *Sky and Telescope,* vol. 23, no. 4, pp. 183, 197, 1962.

Life Supporting Planets, *Sci. News Lett.,* vol. 73, no. 21, p. 322, 1958.

Life Without End, *Time,* vol. 75, p. 54, Jan. 4, 1960.

Lighthall, W. D.: The Law of Cosmic Evolutionary Adaptations, an Interpretation of Recent Thought, *Trans. Royal Soc. Canada,* ser. 3, no. 34, sect. 2, pp. 135–141, 1940.

Lomonaco, T.: The Life in the Universe, *Minerva Med,* vol. 61, pp. 1395–1400 (in Italian), April 7, 1970.

Macvey, John W.: Alone in the Universe?, *Spaceflight,* vol. 4, no. 4, pp. 125–127, 1962.

Man Not in Space, *Sci. News Lett.,* vol. 80, pp. 251–252, Oct. 14, 1961.

Mann, Martin: The Man-Horse From Outer Space, *Popular Science,* vol. 179, no. 4, pp. 19–20, 1961.

Many Habitable Worlds, *Sci. News Lett.,* vol. 47, p. 402, June 30, 1945.

Margaria, Rodolfo: On the Possible Existence of Intelligent Living Beings on Other Planets, *Proc. 12th Int. Astronautical Congress,* Washington, D.C., 1961, pp. 556–563, Academic Press, Inc., New York, 1963; *Revi. Med. Aeron. Spaziale (Roma),* vol. 25, no. 1, pp. 24–35, 1962.

McCarthy, John: Possible Forms of Intelligence: Natural and Artificial, in Cyril Ponnamperuma and A. G. W. Cameron (eds.), "Interstellar Communication: Scientific Perspectives," Houghton Mifflin Company, Boston, 1974.

Me Earthman, You Gasbag, *Newsweek,* vol. 68, pp. 88–89, Aug. 22, 1966.

Moll, H. M.: Bioastronautics and Extraterrestrial Life, NASA TT F-13467, March 1971; translated into English from *Rev. de Aeron. Astronautica* (Madrid), vol. 30, no. 358, pp. 663–673, 1970.

Molton, P. M.: Exobiology, Jupiter, and Life, *Spaceflight,* vol. 14, no. 6, pp. 220–223, 1972.

———: Limitations of Terrestrial Life, *Spaceflight,* vol. 15, no. 1, pp. 27–30, 1973.

———: Terrestrial Biochemistry in Perspective: Some Other Possibilities, *Spaceflight,* vol. 15, no. 4, pp. 139–144, 1973.

Monad, Jacques: "Chance and Necessity: An Essay on the Natural Philosophy of Modern Biology," (trans. from French), Alfred A. Knopf, Inc., New York, 1971.

Moore, Patrick and Francis Jackson: Alien Life, and Summary of Present Position, "Life in the Universe," chaps. 9, 10, pp. 115–128, W. W. Norton and Company, Inc., New York, 1962.

More Evidence for Life in Space, *Sci. Digest,* vol. 70, pp. 30–31, Sept. 1971.

Morrison, Philip: Conclusion: Entropy, Life and Communication, in Cyril Ponnamperuma and A. G. W. Cameron (eds.), "Interstellar Communication: Scientific Perspectives," Houghton Mifflin Company, Boston, 1974.

Moseley, E. L.: Are There Creatures Like Ourselves in Other Worlds?, *Sci. Amer.,* vol. 145, pp. 308–310, 1931.

Motz, Lloyd: Extraterrestrial Intelligence and Stellar Evolution, *Inst. Aerospace Sci. Paper,* no. 63–49, presented at IAS 31st Annual Meeting, New York City, Jan. 21–23, 1963.

Muller, H. J.: Life Forms to be Expected Elsewhere Than on Earth, *Spaceflight,* vol. 5, no. 3, pp. 74–85, 1963.

National Aeronautics and Space Administration: Is There Life In Outer Space?, "Aviation and Cosmonautics," pp. 129–131, NASA TM X-68-82191; Air Force Systems Command, Wright-Patterson AFB, Ohio, Foreign Technology Division, 1966.

Nature's Way, *Time,* vol. 98, pp. 64–65, Nov. 29, 1971.

Noe, Stephen: Correspondence in *Analog,* vol. 86, no. 3, pp. 171–172, 1970.

Notes and News, B. B. C. talk, Is There Life Elsewhere in the Universe? (Fred Hoyle and C. D. Darlington), *J. Brit. Interplanetary Soc.,* vol. 8, no. 5, pp. 200–201, 1949.

Notes and News, The Evolution of Life in the Universe, (report on argument between J. D. Bernal and N. W. Pirie), *J. Brit. Interplanetary Soc.,* vol. 12, no. 4, pp. 180–181, 1953.

O'Brien, Robert: Somebody Up There Like Us, *Esquire,* vol. 60, pp. 185–187, 232, Dec. 1963.

Oparin, A. I.: Genesis of Life on and Beyond the Earth, in A. A. Imshenetskii (ed.), "Extraterrestrial Life and Its Detection Methods," pp. 16–27, 1970.

Ordway, Frederick I., III, James P. Gardner, and Mitchell R. Sharpe, Jr.: Life Outside the Solar System, "Basic Astronautics," Prentice-Hall, Inc., Englewood Cliffs, N. J., 1962.

Osmundsen, John A.: Space-Bug Origin of Life Ruled Out, *New York Times,* Dec. 28, 1961, p. 31.

Ozick, C.: If You Can Read This, You Are Too Far Out, *Esquire,* vol. 79, pp. 74, 78, Jan. 1973.

Pearman, J. P. T.: Extraterrestrial Intelligent Life and Interstellar Communication: An Informal Discussion, in A. G. W. Cameron (ed.), "Interstellar Communication," pp. 287–293, W. A. Benjamin, Inc., New York, 1963.

Pfeiffer, John, Harold F. Blum, Michael Halasz, Leonard Ornstein, and George Gaylord Simpson: Life on Other Planets: Some Exponential Speculations, *Science,* vol. 144, no. 3619, pp. 613–615, 1964.

Phillifent, John T.: Correspondence concerning evolution to unique 'Man,' *Analog,* vol. 85, no. 2, pp. 171–174, 1970.

Pimentel, G. C., K. C. Atwood, Hans Gaffron, H. K. Hartline, T. H. Jukes, E. C. Pollard, and Carl Sagan: Exotic Biochemistry in Exobiology, in C. S. Pittendrigh, Wolf Vishniac, and J. P. T. Pearman (eds.), "Biology and the Exploration of Mars," pp. 243–251, National Academy of Sciences, Washington, D. C., 1966.

Pirie, N. W.: A Discussion on Anomalous Aspects of Biochemistry of Possible Significance in Discussing the Origins and Distribution of Life. Introduction, *Proc. Royal Soc. (Biol.),* vol. 171, pp. 3–4, Aug. 13, 1968.

Ponnamperuma, Cyril: Life in the Universe—Intimations and Implications for Space Science, *Astronautics and Aeronautics,* vol. 3, no. 10, pp. 66–69, 1965.

———: The Search for Extraterrestrial Life, *The Sci. Teacher,* vol. 32, no. 7, pp. 21–26, 1965.

———: A la Recherche de la Vie Extra-terrestre, *Sciences* (Paris), vol. 12, pp. 48–55, 1972.

———: The Chemical Basis of Extraterrestrial Life, in Cyril Ponnamperuma and A. G. W. Cameron (eds.), "Interstellar Communication: Scientific Perspectives," Houghton Mifflin Company, Boston, 1974.

Posin, Dan Q.: Other Suns Other Planets—But Is There Other Life?, *Today's Health,* pp. 56–59, Nov. 1964.

Possible Pattern for the Origin of Life, *Space World,* vol. H-9-93, pp. 47–48, Sept. 1971.

Pryor, H.: Earthmen Are the Only Men, *Sci. Digest,* vol. 55, p. 42, May 1964.

Race of Flying Men, *Southern Rev.,* vol. 10, no. 1832, p. 272.

Rensaw, C. C., Jr. and F. D. Drake: Is There Life Out There?, interview in *National Wildlife,* vol. 10, pp. 50–53, Oct. 1972.

Ribes, J. C. and F. Biraud: Extraterrestrial Civilizations, *Proc. Astronomical Soc. Australia,* vol. 2, no. 1, pp. 11–13, 1971.

Roberts, A. W.: Are the Skies Inhabited? *Sci. Amer. Suppl.,* vol. 59, pp. 24282–24283, Jan. 14, 1905.

Robinson, L.: Are There Men In Other Worlds?, *Cur. Lit.,* vol. 44, pp. 672–677, June 1908.

Rorvik, D. M.: Present Shock, *Esquire,* vol. 78, pp. 134, 136, 138, Dec. 1972.

Russell, H. N.: Are There Other Habitable Worlds?, *Sci. Amer.,* vol. 132, p. 315, 1925.

————: Fading Belief in Life on Other Planets, *Sci. Amer.,* vol. 150, pp. 296–297, June 1934.

Ruzic, Neil P.: "Where the Winds Sleep," Doubleday and Company, Inc., Garden City, N. Y., 1970; pertinent section serialized in *Industrial Research,* part 5, pp. 79–80, May 1970.

Sagan, Carl and Norton Leonard: The Solar System and Beyond, "Planets: Life Science Library," chap. 8, pp. 176–191, Time-Life Books, New York, 1966.

————: Interstellar Organic Chemistry, *Nature,* vol. 238, pp. 77–80, July 14, 1972.

————: Life Beyond the Solar System, in Cyril Ponnamperuma (ed.), "Exobiology," North-Holland Publishing Co., Amsterdam, 1972.

Schmeck, Harold M., Jr.: Life Held Likely on Other Worlds, *New York Times,* p. 29, March 2, 1961.

————: Space 'Centaurs' Thought to Exist, *New York Times,* p. 21, July 4, 1961.

Serdobolskii, V. I.: The Problem of the Distribution of Intelligent Life in the Universe, in A. I. Ishlinskii and V. A. Petukov (eds.), "Physics, Mathematics, Mechanics," pp. 145–154 (in Russian), Proc. of the 1st Moscow Conference of Young Scientists, 1968.

Shapley, Harlow: Extraterrestrial Life, *Astronautics,* vol. 5, pp. 32–33, 50, April 1960.

————: Life on Unseen Planets, *Sci. Digest,* vol. 53, pp. 59–64, Feb. 1963; *The Am. Scholar,* 1962.

Silicon-Based Life Possible, *Sci. News Lett.,* vol. 78, no. 22, p. 342, 1960.

Simpson, George Gaylord: The Nonprevalence of Humanoids, *Science,* vol. 143, no. 3608, pp. 769–775, 1964.

————: "This View of Life: The World of an Evolutionist," Harcourt, Brace, and World, Inc., New York, 1964.

————: Space Flights and Biology, correspondence in *Science,* vol. 144, no. 3616, p. 246, 1964.

Slater, A. E.: The Probability of Intelligent Life Evolving on a Planet, *Proc. 8th Int. Astronautical Congress,* Barcelona, 1957, pp. 395–402, Springer-Verlag OHG, Vienna, 1958.

————: "Life in the Universe," *Spaceflight,* vol. 4, p. 88, May 1962.

————: Extra-Terrestrial Life, correspondence in *Spaceflight,* vol. 5, no. 1, p. 36, 1963.

————: Nucleic Acid and the Improbability of Life, *Proc. 16th Int. Astronautical Congress: Life in Spacecraft,* Athens, 1965, pp. 165–170, Gordon and Breach, New York, 1966.

Space Men Predicted, *Sci. News Lett.,* vol. 80, p. 35, July 15, 1961.

Spall, N. J.: Will 'They' Look Like Us?, correspondence in *Spaceflight,* vol. 14, no. 11, p. 437, 1972.

Spontaneous Life Theory Advanced by Biologist, *Sci. News Lett.,* vol. 77, no. 10, p. 150, 1960.

Struve, Otto: Life on Other Worlds, *Sky and Telescope,* vol. 14, no. 4, pp. 137–140, p. 146, 1955.

Stuhlinger, Ernst: Life on Other Stars: Parts I and II, *Space J.,* vol. 1, no. 2, pp. 10–16, 1958; vol. 1, no. 3, pp. 21–30, 1958.

Sullivan, Walter: Milky Way Yields a Clue About Life, *New York Times,* p. 43, March 21, 1969.

———: Scientists Find Alcohol Cloud Near Center of the Milky Way, *New York Times,* p. 57, Nov. 6, 1970.

———: Origin of Life: It May Have Been Far Out in Space, *New York Times,* sect. 4, p. 14, Nov. 8, 1970.

Suspect 'Human' Life on Millions of Planets, *Sci. News Lett.,* vol. 74, no. 21, p. 328, 1958.

Techtran Corp., Glen Burnie, Md.: Life in Space—New Advances in Space Biology, (translated from the German), NASA TT F-11540, Feb. 1968.

The Transmission of Life to a Dead World Despite the Ultraviolet Ray, *Curr. Opinion,* vol. 56, p. 286, April 1914.

Turner, B. E.: Interstellar Molecules, *Sci. Amer.,* vol. 228, pp. 50–62, 67–69, March 1973.

Underwood, R. S.: Are We Alone In the Universe?, *Sci. Monthly,* vol. 49, pp. 155–159, Aug. 1939.

Uphof, J. C. T.: The Dynamics of the Distribution of Life in the Universe, *Scientia,* vol. 58, pp. 30–38, July 1935.

Vertregt, M.: Cosmic Squeak and Gibber, *Spaceflight,* vol. 7, no. 4, pp. 122–128, 1965.

Vishniac, Wolf: Space Flights and Biology, correspondence in *Science,* vol. 144, no. 3616, pp. 245–246, 1964.

Voronin, M. A.: In Quest of Signs of Civilization in Other Worlds, (trans. from Russian), *Priroda,* no. 11, pp. 78–83, 1962, NASA TT F-8590.

Waterless Life, *Time,* vol. 98, p. 65, Nov. 29, 1971.

Webster, Gary: Is There Life on Other Worlds?, *Natural History,* vol. 65, pp. 526–531, Dec. 1956.

Whitrow, G. J.: Mutation Rates and Stellar Explosions, in I. J. Good (ed.), "The Scientist Speculates," pp. 233–235, William Heinemann, Ltd., London, 1962.

Zuckerman, B.: Interstellar Molecules and Interstellar Life, *N. Y. S. J. Med.,* vol. 70, p. 50, 1970.

II.   Methods of Communicating with Extrasolar Intelligence

A.   *Searching for Signals from Extrasolar Intelligence*

Anderson, Leland I.: Extra-Terrestrial Radio Transmissions, *Nature*, vol. 190, p. 374, April 22, 1961.

Anybody Out There?, *Time*, vol. 74, no. 21, pp. 84–85, 1959.

Ashbrook, Joseph: Astronomical Scrapbook—Beginnings of the Space Age, *Sky and Telescope*, vol. 22, no. 2, p. 85, 1961.

Asimov, Isaac: Hello, CTA-21—Is Anyone There?, *New York Times Magazine*, Nov. 29, 1964.

———: Microwave Radio Beams from Space Might Be Signals from Other Intelligences, *Catholic Digest*, vol. 29, pp. 37–40, June 1965.

———: Is Anyone There?, "Is Anyone There?," Doubleday and Company, Inc., Garden City, N. Y., pp. 197–206, 1967.

Ball, Gloria: Listen in on Other Suns, *Sci. News Lett.*, vol. 77, pp. 282–283, April 30, 1960.

Barrett, Alan H.: Radio Observations of Interstellar Hydroxyl Radicals, *Science*, vol. 157, no. 3791, pp. 881–889, 1967.

Bart, Peter: Symposium on Coast Assays Life On Other Planets, *New York Times*, May 29, 1966, p. 58.

———: Scientist Tells Meeting at Caltech He is Doubtful of Interstellar Messages, *New York Times*, p. 37, Oct. 26, 1966.

Bayley, Donald S.: The Interplanetary and Interstellar Communication Potential of the Laser, NAECON Proceedings, 1962.

Belitsky, Boris: Signals from Other Worlds, *Spaceflight*, vol. 14, no. 1, pp. 17–18, 1972.

Beller, William: How to Contact 'People' in Space?, *Missiles and Rockets*, vol. 7, pp. 42–44, July 25, 1960.

Bergier, Jacques: *A L'Écoute des Planètes*, Fayard, Paris, 1963.

Big Radio Ear Listens for Planets' Signals, *New York Times*, April 5, 1960, p. 13.

Binder, Otto O.: Star Search, *Space World*, pp. 27, 50–51, July 1962.

Boehm, G. A. W.: Are We Being Hailed from Interstellar Space?, *Fortune*, vol. 63, pp. 144–149, 193–194, March 1961.

Boni, A.: Telecomunicazioni spaziali e caratteristiche dei loro canali, (Telecommunication Systems in Space and Properties of their Channels), *Proc. 10th Int. Astronautical Congress*, London, 1959, pp. 871–910, Springer-Verlag OHG, Vienna, 1960.

———: Stellar Signals Program, paper abstract in *Proc. 11th Int. Astronautical Congress*, Stockholm, 1960, p. 306, Springer-Verlag OHG, Vienna, 1961.

Bova, Ben: It's Right Over Your Nose, *Analog*, vol. 81, no. 4, pp. 95–100, 1968.

Bracewell, R. N.: Communications from Superior Galactic Communities, in A. G. W. Cameron (ed.), "Interstellar Communication," pp. 242–248,

W. A. Benjamin, Inc., New York, 1963; *Nature,* vol. 186, no. 4726, pp. 670–671, 1960.

———: Radio Signals from Other Planets, in A. G. W. Cameron (ed.), "Interstellar Communication," pp. 199–200, W. A. Benjamin, Inc., New York, 1963; *IRE, Proc.,* vol. 50, p. 214, Feb. 1962.

Bradbury, Ray: A Serious Search for Weird Worlds, *Life,* vol. 49, no. 17, pp. 116–118, 120, 123–124, 126, 128, 130, 1960.

Briggs, Michael H.: Superior Galactic Communities, *Spaceflight,* vol. 3, pp. 109–110, May 1961.

Budden, K. G. and G. G. Yates: A Search for Radio Echoes of Long Delay, *J. Atmospheric and Terrestrial Phys.,* vol. 2, pp. 272–281, 1952.

Butler, Clay P.: The Light of the Atom Bomb, *Science,* vol. 138, no. 3539, pp. 483–489, 1962.

Cade, C. M.: Communicating with Life in Space, *Discovery,* vol. 24, pp. 36–41, May 1963.

Call Signs from Space, *Spaceflight,* vol. 7, no. 5, pp. 165–167, 1965.

Calvin, M.: Talking to Life on Other Worlds, *Sci. Digest,* vol. 53, pp. 14–19, 88, 89, Jan. 1963.

Cameron, A. G. W.: Communicating with Intelligent Life on Other Worlds, *Sky and Telescope,* vol. 26, no. 5, p. 258, 1963.

———: Future Research on Interstellar Communication, in A. G. W. Cameron (ed.), "Interstellar Communication," pp. 309–315, W. A. Benjamin, Inc., New York, 1963.

CETI Questionnaire, responses from Soviet specialists, *Spaceflight,* vol. 15, no. 4, pp. 137–138, 1973; translated from the Russian, *Zemlya i Vselennaya,* 1972.

Chances of Contacting Extraterrestrial Civilizations Seem Poor, *Sci. News,* vol. 103, p. 118, Feb. 24, 1973.

Clark, A. C.: Messages from the Invisible Universe, in Richard M. Skinner and William Leavitt (eds.), "Speaking of Space: The Best from Space Digest," pp. 32–39, Little, Brown and Company, Boston, 1962.

Cocconi, Giuseppe and Philip Morrison: Searching for Interstellar Communications, in A. G. W. Cameron (ed.), "Interstellar Communication," pp. 160–164, W. A. Benjamin, Inc., New York, 1963; *Nature,* vol. 184, pp. 844–846, 1959.

——— and ———: Problems of Interstellar Communication, in V. I. Krasovskiy (ed.), "Space—Collection of Articles," (trans. from Russian, *Kosmos* (Moscow), no. 1, pp. 82–85, 1963), Air Force Systems Command, Wright-Patterson AFB, Ohio, Foreign Technology Division, FTD-MT-64-239, AD-608184.

Corliss, William R.: The Search for Life Beyond the Earth, "Mysteries of the Universe," chap. 11, pp. 180–207, Thomas Y. Crowell Company, New York, 1967.

Drake, Frank D.: How Can We Detect Radio Transmissions from Distant

Planetary Systems?, in A. G. W. Cameron (ed.), "Interstellar Communication," pp. 165–175, W. A. Benjamin, Inc., New York, 1963; *Sky and Telescope,* vol. 19, no. 3, pp. 140–143, 1960.

———: "Project Ozma," in A. G. W. Cameron (ed.), "Interstellar Communication," pp. 176–177, W. A. Benjamin, Inc., New York, 1963; "McGraw-Hill Yearbook of Science and Technology," 1962; *Phys. Today,* vol. 14, pp. 40–42, 44, 46, April 1961.

———: InterGalactic Communication, discussion in Boston; summarized in *PGNE News, IEEE Trans. on Aerospace and Navigational Electronics;* Sept. 1962, Dec. 1962, March 1963.

———: The Radio Search for Intelligent Extraterrestrial Life, in G. Mamikunian and M. H. Briggs (eds.), "Current Aspects of Exobiology," pp. 323–345, Pergamon Press, New York, 1965.

———: Prospects in the Search for Extraterrestrial Civilizations, 12th Ann. Meeting of the Amer. Astronautical Soc., Anaheim, Calif., May 23–25, 1966, Paper 66–78.

———: Intelligent Life in Other Parts of the Universe, in Hugh Odishaw (ed.), "The Earth in Space," chap. 30, pp. 308–316, Basic Books, Inc., Publishers, New York, 1967.

———: Methods of Communication: Message Content; Search Strategy; Interstellar Travel, in Cyril Ponnamperuma and A. G. W. Cameron (eds.), "Interstellar Communication: Scientific Perspectives," chap. 8, Houghton Mifflin Company, Boston, 1974.

Dugan, George: Other-World Bid to Earth Doubted, *New York Times,* p. 31, June 24, 1960.

DuShane, G.: Next Question, *Science,* vol. 130, p. 1733, Dec. 25, 1959.

———: Hello Out There, *New Republic,* vol. 142, p. 9, Jan. 25, 1960.

Dyson, Freeman J.: Search for Artificial Stellar Sources of Infrared Radiation, in A. G. W. Cameron (ed.), "Interstellar Communication," pp. 111–114, W. A. Benjamin, Inc., New York, 1963; *Science,* vol. 131, pp. 1667–1668, June 3, 1960.

———: The Search for Extraterrestrial Technology, in R. E. Marshak (ed.), "Perspectives in Modern Physics, Essays in Honor of Hans A. Bethe," pp. 641–655, John Wiley and Sons, Inc., New York, 1966.

———: Letter from Armenia, *The New Yorker,* Nov. 6, 1971, pp. 126–137.

Earth to 'Listen' for Other World, *New York Times,* p. 30, April 3, 1960.

'Ear' Toward Space, *New York Times,* sect. 4, p. 7, April 17, 1960.

Edie, Leslie C.: Messages from Other Worlds, correspondence in *Science,* vol. 136, no. 3511, p. 184, 1962.

Edson, J. B.: Tuning In On Other Worlds, *New York Times,* sect. 6, pp. 31, 78–79, March 13, 1960.

Elsnau, Mary: Is There Intelligent Life on the Planets?, *Amer. Mercury,* vol. 91, no. 441, pp. 32–45, 1960.

Filipowsky, R. F. and E. I. Muehldorf: Stellar Communications, "Space Communications Systems," pp. 482–492, Prentice-Hall, Inc., Englewood Cliffs, N.J., 1965.

Finney, John W.: Radio Ear Cocked for Talk in Space, *New York Times*, p. 3, April 13, 1960.

———: Scientists and Congress Ponder If Life Exists in Other Worlds, *New York Times*, pp. 1, 16, March 23, 1962.

First Soviet-American Conference on Communication with Extraterrestrial Intelligence (CETI), *Bioscience*, vol. 21, no. 23, 1971, 1177.

First Soviet-American Conference on Communication with Extraterrestrial Intelligence (CETI), *Icarus*, vol. 16, pp. 412–414, 1972.

First Soviet-American Conference on Communication with Extraterrestrial Intelligence (CETI), *Spaceflight*, vol. 14, no. 1, pp. 18–19, 1972.

Fitch, C. J.: Interplanetary Communication, *Astronautics*, vol. 4, pp. 1–2, 5–7, Oct. 1930.

Flammarion, C.: Inter-Astral Communication, *Amer. Rev. of Reviews*, vol. 5, p. 90, Feb. 1892.

Forkosch, Morris D.: Reaching Other Galaxies, correspondence in *New York Times*, Feb. 14, 1962, p. 34.

Frisch, Bruce: Dyson's Cool Worlds, *Sci. Digest*, vol. 63, no. 6, pp. 35–41, 1968.

Galton, F.: Intelligible Signals Between Neighboring Stars, *Fortnightly Rev.* (London), vol. 66, pp. 657–664, Nov. 1896.

Gardner, Martin: Mathematical Games: Communications with Intelligent Organisms on Other Worlds, *Sci. Amer.*, vol. 213, pp. 96–100, Aug. 1965.

Geiger, Richard: correspondence in *Analog*, vol. 82, no. 4, pp. 167–168, 1968.

Gindilis, L. M.: Extraterrestrial Civilizations—Subject of Explorations and Investigations, *Zemlya i Vselennaya*, no. 5, pp. 2–7, 9 (in Russian), 1970.

———: The Possibility of Radio Communication with Extraterrestrial Civilizations, in S. A. Kaplan (ed.), "Extraterrestrial Civilizations," chap. 3, NASA TT F-631, (trans. from Russian) 1971.

———: Problems of Communication with Extraterrestrial Intelligence (CETI)—Symposium in Byurakan, *Vestnik Akademii Nauk SSSR*, vol. 1972, p. 82, 1972.

Golay, M. J. E.: Coherence in Interstellar Signals, in A. G. W. Cameron (ed.), "Interstellar Communication," pp. 192–198, W. A. Benjamin, Inc., New York, 1963; *IRE, Proc.*, vol. 49, pp. 958–959, 1961; vol. 50, p. 223, 1962.

Golomb, Solomon W.: New Problems of Space Communications: Part 1—Beware of Tigers, *Astronautics*, vol. 7, no. 6, p. 19, 1962.

Graham, E. A.: A System Design for Improved Extra-Terrestrial Communications, *IEEE Transactions on Aerospace and Electronics Systems*, vol.

AES-3, pp. 803–807, 1967; "Record of the 1965 International Symposium in Space Electronics," Miami Beach, Fla., pp. 12-D1 to 12-D5, Nov. 24, 1965.

Gudzenko, L. I. and B. N. Panovkin: Reception of Signals Transmitted by Extraterrestrial Civilizations, in G. M. Tovmasyan (ed.), "Extraterrestrial Civilizations," NASA Scientific and Technical Information Facility translation, N67-30335, pp. 43–45.

Guillemin, A. V.: Communication With the Planets, *Popular Sci.,* vol. 40, pp. 361–363, Jan. 1892.

Habitable Shells Pictured in Stars, *New York Times,* p. 11, June 3, 1960.

Hafner, Everett M.: Galactic Signals, *Bull. Atomic Scientists,* vol. 23, pp. 50–52, June 1967.

———: Techniques of Interstellar Communication, in M. M. Freundlich and B. M. Wagner (eds.), "Exobiology," *AAS Science and Technology Series,* vol. 19, pp. 37–62, 1967.

Handelsman, Morris: "Considerations on Communication with Intelligent Life in Outer Space," 1962 WESCON Convention Record, part 5, paper 4.4, Los Angeles, Calif., Aug. 21–24, 1962.

Haviland, R. P.: On the Search for Extra-Solar Intelligence, *Spaceflight,* vol. 14, no. 6, pp. 217–219, 223, 1972.

Herbert, John D.: Man and Amoeba, correspondence in *New York Times,* Jan. 21, 1962, sect. 6, p. 4.

Hicks, Clifford B.: We're Listening for Other Worlds, *Popular Mechanics,* vol. 114, no. 3, pp. 81–85, 220, 222, 224, 226, 1960.

von Hoerner, Sebastian: The Search for Signals from Other Civilizations, in A. G. W. Cameron (ed.), "Interstellar Communication," pp. 272–286, W. A. Benjamin, Inc., New York, 1963; *Science,* vol. 134, pp. 1839–1843, Dec. 1961.

Hoffleit, Dorrit: Communication with Other Planets!, in Thornton Page and Lou William Page (eds.), "The Origin of the Solar System," pp. 293–294, The Macmillan Company, New York, 1966.

Holden, C.: Soviet-American Conference Urges Search for Other Worlds, *Science,* vol. 174, no. 4005, pp. 130–131, 1971.

Horowitz, N. H., F. D. Drake, S. L. Miller, L. E. Orgel, and C. Sagan: The Origins of Life, in P. Handler (ed.), "Biology and the Future of Man," Oxford Press, New York, 1970.

Huang, Su-Shu: "Problems of Transmission in Interstellar Communication," unclassified NASA report available to U. S. government agencies and contractors only, X63-11603, 1962.

———: "Problem of Transmission in Interstellar Communication," in A. G. W. Cameron (ed.), "Interstellar Communication," pp. 201–206, W. A. Benjamin, Inc., New York, 1963.

International Academy of Astronautics of the International Astronautical Federation, International Review Meeting on Communication with Extraterrestrial Intelligence, Vienna, Oct. 8–15, 1972, R. Pesek (chairman), proceedings to be published.

Is Anybody Out There Sending?, *Sci. News,* vol. 100, no. 14, pp. 223–224, 1971.

Jackson, A. A.: Signs of Extraterrestrial Civilization, correspondence in *Analog,* vol. 83, no. 6, p. 175, 1969.

Jackson, C. D. and R. E. Hohmann: "An Historic Report on Life in Space: Tesla, Marconi, Todd," 17th Annual Meeting of the American Rocket Society and Space Flight Exposition, Los Angeles, Calif., Nov. 13–18, 1962, paper 2730-62.

Jarnagin, William S.: Intragalactic Communication, correspondence in *Science,* vol. 131, no. 3408, pp. 1262–1263, 1960.

Kaplan, S. A.: Exosociology—The Search for Signals from Extraterrestrial Civilizations, in S. A. Kaplan (ed.), "Extraterrestrial Civilizations," pp. 1–12, NASA TT F-631, (trans. from Russian), 1971.

Kardashev, N. S.: Transmission of Information by Extraterrestrial Civilizations, in G. M. Tovmasyan (ed.), "Extraterrestrial Civilizations," NASA Scientific and Technical Information Facility translation, N67-30332, pp. 19–29.

———: Transmission of Information by Extraterrestrial Civilizations, *Soviet Astronomy,* vol. 8, no. 2, pp. 217–221, 1964; *Astron. Zh.* (Moscow), vol. 21, no. 2, pp. 282–287, 1964.

———: The Astrophysical Aspects of the Search for Signals from Extraterrestrial Civilizations, in S. A. Kaplan (ed.), "Extraterrestrial Civilizations," (trans. from Russian), NASA TT F-631, 1971.

Khaikin, S. E.: Communication with Extraterrestrial Civilizations, in G. M. Tovmasyan (ed.), "Extraterrestrial Civilizations," NASA Scientific and Technical Information Facility translation, N67-30336, pp. 53–59.

Koestler, Arthur: Trying to Talk with Other Planets, *Sci. Digest,* vol. 47, p. 69, June 1960.

Kotelnikov, V. A.: Radio Communication with Extraterrestrial Civilizations, in G. M. Tovmasyan (ed.), "Extraterrestrial Civilizations," NASA Scientific and Technical Information Facility translation, N67-30339, pp. 72–77.

Kross, Robert D.: Space Messengers, correspondence in *Science,* vol. 136, no. 3519, pp. 913–914, 1962.

Kulin, D.: How Man Might Communicate with Other Planets, "Problems of Space Research Investigated," (trans. from Russian, *Nauk i Zhizn,* no. 3, 1968), Joint Publications Research Service, JPRS-46446, pp. 13–18.

Kuznetsov, Yu. P.: Signals of Extraterrestrial Civilizations, *USSR Sci. Abstracts: Electronics and Electrical Eng.,* no. 133 (Microfilm Reel R357, Microfiche No. R11047), *RZh-Radiotekhnika,* Jan. 1971.

———: Signals of Extraterrestrial Civilizations—Of What Kind Can They

Be?, *Zemlya i Vselennaya,* no. 1, pp. 30–33 (in Russian), 1972.

Lapp, Ralph E.: How to Talk to People, If Any, on Other Planets, *Harper's Magazine,* vol. 222, pp. 58–63, March 1961.

Lawrence, William L.: Radio Astronomers Listen for Signs of Life in Distant Solar Systems, *New York Times,* Nov. 22, 1959, sect. 4, p. 11.

Lawrence, L. George: Interstellar Communications—What are the Prospects?, *Electronics World,* vol. 86, no. 4, pp. 34–35, 56, 1971.

Lawton, A. T.: Infra-red Interstellar Communication, *Spaceflight,* vol. 13, no. 3, pp. 83–85, 1971.

———: Startalk—The Problems of Interstellar Communication, *Spaceflight,* vol. 13, no. 7, pp. 241–244, 1971.

Lear, John: The Search for Intelligent Life on Other Planets, *Sat. Rev.,* vol. 43, pp. 39–43, Jan. 2, 1960.

———: The Search for Man's Relatives Among the Stars, *Sat. Rev.,* pp. 29–37, June 10, 1972.

Ley, Willy: Epilogue: The Search for Other Civilizations, "Watchers of the Skies," pp. 501–504, The Viking Press, Inc., New York, 1963.

Life Beyond Earth Sought: Project Ozma, *Sci. News Lett.,* vol. 84, p. 166, Sept. 14, 1963.

Listening for Life on Other Worlds, *New York Times,* Oct. 8, 1961, sect. 12, p. 9.

Lovell, Bernard: Search for Voices from Other Worlds, *New York Times,* Dec. 24, 1961, sect. 6, pp. 10, 29–31.

Lust for Life, *Sci. News,* vol. 93, no. 2, pp. 29–31, 1968.

Luyten, W. J. and Sebastian von Hoerner: The Search for Other Civilizations, correspondence in *Science,* vol. 135, no. 3507, pp. 991–992, 994, 1962.

Macvey, John W.: Interstellar Beacons, *Spaceflight,* vol. 14, no. 1, pp. 14–16, 25, 1972.

Marx, G.: "Messages from Outer Space," ASTIA Report No. AD 270779, Defense Documentation Center, Washington, D.C. (translation from *Tizikai Szemele,* vol. 10, no. 11, p. 335, 1960).

Maser Here. Hello There!, *Newsweek,* vol. 70, p. 84, Sept. 11, 1967.

McDonough, Thomas R.: They're Trying to Tell Us Something, *Analog,* vol. 83, no. 1, pp. 62–79, 1969 (Part I); and vol. 83, no. 2, pp. 74–86, April 1969 (Part II).

Morrison, Philip: Extraterrestrial Contact, in Jerry Grey and Vivian Grey (eds.), "Space Flight Report to the Nation," ch. 16, pp. 144–158, Basic Books, Inc., Publishers, New York, 1962.

———: Interstellar Communication, in A. G. W. Cameron (ed.), "Interstellar Communication," pp. 249–271, W. A. Benjamin, Inc., New York, 1963; *Bull. Phil. Soc. Washington,* vol. 16, p. 58, 1962.

———: Outlook Regarding Interstellar Communication, in A. G. W. Cameron (ed.), "Interstellar Communication," pp. 316–317, W. A. Benjamin, Inc., New York, 1963.

———: Listening for Life on Other Worlds, *Boston Evening Globe,* Sept. 29, 1971; But It Was No Science Fiction Convention, *Tech Talk—M.I.T.,* vol. 16, no. 12, Sept. 1971.

Mukhin, L. M.: "Current Position on CETI from the Viewpoint of Biology," paper in Russian at the International Astronautical Federation, 23rd Int. Astronautical Congress, Vienna, Oct. 8–15, 1972.

Mutschall, Vladimir: Interstellar Communications, *Foreign Sci. Bull.,* vol. 1, no. 9, pp. 56–63, Library of Congress, Washington, D.C., Aerospace Technology Division, 1965.

National Academy of Sciences-Astronomy Survey Committee, "Astronomy and Astrophysics for the 1970's," 1972.

New Evidence of Intelligent Life on Other Worlds, *Sci. Digest,* vol. 57, pp. 8–10, Feb. 1965.

A New Look at Space and Time, editorial in *New York Times,* Feb. 5, 1962, p. 30.

"Nobody There," *Newsweek,* vol. 71, p. 63, Apr. 22, 1968.

Oliver, B. M.: Radio Search for Distant Races, *Int. Sci. and Tech.,* pp. 55–60, 96, Oct. 1962.

———: First Picture from Another Planet, *Student Quart. and EE Digest,* pp. 48–53, Sept. 1962.

———: Some Potentialities of Optical Masers, in A. G. W. Cameron (ed.), "Interstellar Communication," pp. 207–222, W. A. Benjamin, Inc., New York, 1963; *IRE, Proc.,* vol. 50, pp. 135–141, 1962.

———: Interstellar Communication, in A. G. W. Cameron (ed.), "Interstellar Communication," pp. 294–305, W. A. Benjamin, Inc., New York, 1963; *1962 IRE International Convention Record,* vol. 10, part 8, pp. 34–39, March 26–29, 1962.

———: Technical Considerations in Interstellar Communication, in Cyril Ponnamperuma and A. G. W. Cameron (eds.), "Interstellar Communication: Scientific Perspectives," chap. 9, Houghton Mifflin Company, Boston, 1974.

——— and John Billingham: "Project Cyclops: A Design Study of a System for Detecting Extraterrestrial Intelligent Life," prepared under Stanford/NASA/Ames Research Center 1971 Summer Faculty Fellowship Program in Engineering System Design, NASA CR 114445; International Astronautical Federation, 23rd Int. Astronautical Congress, Vienna, Oct. 8–15, 1972.

Ozmology, *Sci. Amer.,* vol. 203, no. 5, pp. 97–98, 1960.

Page, Thornton and Lou Williams Page: Signaling the Inhabitants of Other Planets, "The Origin of the Solar System," pp. 292–293, Macmillan Company, New York, 1966.

Panovkin, B. N.: Extraterrestrial Civilizations and Kybernetics, *Zemlya i Vselennaya,* no. 6, pp. 2–5 (in Russian), 1969.

———: Extraterrestrial Civilizations—Problems and Considerations, *Priroda,* no. 7, pp. 56–61 (in Russian), 1971.

———: The Effect of the Space Medium on the Propagation of Radio Signals, in S. A. Kaplan (ed.), "Extraterrestrial Civilizations," pp. 59–68, NASA TT F-631, trans. from Russian, 1971.

Pariiskii, Yu. N.: Observations of the Peculiar Radio Sources CTA 21 and CTA 102 at Pulkovo, in G. M. Tovmasyan (ed.), "Extraterrestrial Civilizations," NASA Scientific and Technical Information Facility translation, N67-30333, pp. 30–37.

Pokrovsky, Georgi: Where Should Space Neighbors Be Looked For?, *Space World,* vol. 1-7-103, pp. 46–47, 1972; vol. 1-10-106, pp. 15–16, 1972.

van der Pol, Balth: Short Wave Echoes and the Aurora Borealis, *Nature,* vol. 122, no. 3084, pp. 878–879, 1928.

Price, George R.: U.S. Begins Search for Beings in Other Worlds, *Popular Sci.,* vol. 176, pp. 66–69, 209, April 1960.

Project Ozma, *Time,* vol. 75, p. 53, Apr. 18, 1960.

Project Ozma Off, *Senior Scholastic* (teacher's edition), vol. 78, p. 20, April 26, 1961.

Rosenberg, Paul: Communication with Extraterrestrial Intelligence, *Aerospace Eng.,* vol. 21, no. 8, pp. 68–69, 111, 1962.

Ross, Monte: Search Via Laser Receivers for Interstellar Communications, *IEEE, Proc.,* vol. 53, p. 1780, Nov. 1965.

———: Interstellar Communication with Others, "Laser Receivers," pp. 383–385, John Wiley and Sons, Inc., New York, 1966.

Rublowsky, John: Project Ozma, "Is Anybody Out There?," Walker Publishing Company, New York, 1962.

Sagan, Carl and Russell G. Walker: The Infrared Detectability of Dyson Civilizations, *Astrophysical J.,* vol. 144, no. 3, pp. 1216–1218, 1966.

———: OH Emission Regions and Extraterrestrial Intelligence, *Astrophys. and Space Sci.,* vol. 1, p. 273, 1968.

———: The Search for Extraterrestrial Intelligence, in Cyril Ponnamperuma (ed.), "Exobiology," North-Holland Publishing Co., Amsterdam, 1972.

Schwartz, R. N. and C. H. Townes: Interstellar and Interplanetary Communication by Optical Masers, in A. G. W. Cameron (ed.), "Interstellar Communication," pp. 223–231, W. A. Benjamin, Inc., New York, 1963; *Nature,* vol. 190, pp. 205–208, 1961.

Search for Civilizations, *Sci. News Lett.,* vol. 80, p. 414, Dec. 23, 1961.

Seeking Extraterrestrial Civilizations, *Spaceflight,* vol. 13, no. 12, p. 464, 1971.

Shells Around Suns May Have Been Built, *Sci. News Lett.,* vol. 77, p. 389, June 18, 1960.

Shklovskii, I. S.: Is Communication Possible with Intelligent Beings on Other Planets?, in A. G. W. Cameron (ed.), "Interstellar Communication," pp. 5–16, W. A. Benjamin, Inc., New York, 1963; *Priroda,* no. 7, p. 21 (in Russian), 1960; trans. from Russian, Report M-MS-IS-61-2, George C. Marshall Space Flight Center, Huntsville, Ala.

———: Multiplicity of Inhabited Worlds and the Problem of Interstellar Communications, in G. M. Tovmasyan (ed.), "Extraterrestrial Civilizations," NASA Scientific and Technical Information Facility translation, N67-30331, pp. 5–13.

Siforov, V. I.: Some Aspects of the Search for Signals from Other Civilizations and Their Analysis, in G. M. Tovmasyan (ed.), "Extraterrestrial Civilizations," NASA Scientific and Technical Information Facility translation, N67-30340, pp. 78–83.

Signal from Space?, editorial in *New York Times,* April 13, 1965, p. 36.

Signals from Space?, *Sci. News Lett.,* vol. 79, p. 295, May 13, 1961.

Simak, Clifford D.: Life and Intelligence in the Universe, "Wonder and Glory: The Story of the Universe," chap. 17, pp. 209–232, St. Martin's Press, Inc., New York, 1969.

Slysh, V. I.: Radio Astronomic Artificiality Criteria of Radio Sources, in G. M. Tovmasyan (ed.), "Extraterrestrial Civilizations," NASA Scientific and Technical Information Facility translation, N67-30334, pp. 38–42.

Smirnova, N. A. and N. L. Kaidanovskii: The Influence of the Space Environment and of the Earth's Atmosphere on the Apparent Angular Size of Radio Sources, in G. M. Tovmasyan (ed.), "Extraterrestrial Civilizations," NASA Scientific and Technical Information Facility translation, N67-30341, pp. 84–87. All-Union Conference, 1st, Yerevan, Armenian SSR, May 20–23, 1964, Proc.; Yerevan, Izdatel'stvo Akademii Nauk Armianskoi SSR, 1965, 129–135; discussion, 136–142 (in Russian).

Soviet Scientists Propose an International Program, *New York Times,* Feb. 3, 1966, p. 2.

Soviet Writers Say Earth Has Received Signals from Space, *New York Times,* March 21, 1964, p. 4.

Spacious Talk, *Sci. Amer.,* vol. 202, no. 1, pp. 74, 76, 79, 1960.

Stockton, Bill: Calling All Stars, *Los Angeles Times, West Magazine,* Sept. 10, 1972, pp. 9–12, 14.

Störmer, Carl: Short Wave Echoes and the Aurora Borealis, *Nature,* vol. 122, no. 3079, p. 681, 1928.

The Strange Intruder, *Newsweek,* vol. 56, p. 83, July 4, 1960.

Strickling, H. L.: Any Survivors?, correspondence in *New York Times,* Jan. 28, 1962, sect. 6, p. 19.

Struve, Otto: Detecting Radio Signals from Outside Solar System, correspondence in *New York Times,* Dec. 6, 1959, sect. 4, p. 10.

———: Project OZMA, in Astronomers in Turmoil, *Phys. Today,* vol. 13, no. 9, pp. 22–23, 1960.

———: et al.: correspondence concerning Ray Bradbury's "Search for

Weird Worlds," *Life,* vol. 49, no. 20, p. 22, 1960.

Sullivan, Walter: A Satellite from Another World May Be in Orbit, Scientist Says, *New York Times,* June 20, 1960, pp. 1, 4.

———: Contact with Worlds in Space Explored By Leading Scientists, *New York Times,* Feb. 4, 1962, pp. 1, 44.

———: Search for Life in Nebula Urged, *New York Times,* Dec. 3, 1962, p. 36.

———: Space Signals Said to Hint Life Afar, *New York Times,* Oct. 26, 1964, pp. 1, 36.

———: Communication with Other Planets, *Saturday Review,* vol. 47, pp. 78–79, Dec. 5, 1964.

———: Television from Other Worlds Called Detectable, *New York Times,* Jan. 2, 1966, p. 14.

———: Soviet Urges Search for Signals from Space, *New York Times,* May 10, 1966, p. 19.

———: Space Waves Spur Search for Signals, *New York Times,* Aug. 26, 1967, pp. 1, 10.

———: Astronomers Hear Signals from Space, *New York Times,* March 10, 1968, pp. 1, 92.

———: 'Spooky' Signals from Space, *New York Times,* March 17, 1968, sect. 4, p. 11.

———: British Astronomer Describes 3 More Sources of Radio Signals in Space, *New York Times,* March 29, 1968, p. 19.

———: Listening to Signals in Space, *New York Times,* April 4, 1968, pp. 49, 51.

———: Neutron Stars May Be the Origin of Signals, *New York Times,* April 12, 1968, p. 16.

———: Pulsations from Space, *New York Times,* April 14, 1968, sect. 4, p. 18.

———: The Universe Is Not Ours Alone, *New York Times Magazine,* Sept. 29, 1968, pp. 40–41 *ff.*

———: NASA Seeks 'Ear' to Distant Civilization, *New York Times,* June 16, 1971, p. 13.

———: Man Listens for Life on Worlds Afar, *New York Times,* Sept. 13, 1971, p. 1.

———: Is Someone Out There Trying to Tell Us Something?, *New York Times,* Sept. 19, 1971, News of the Week in Review, p. 8.

Tanner, Henry: Russians Temper Report on Space, *New York Times,* April 14, 1965, p. 3.

——— and Walter Sullivan: Russians Say A Cosmic Emission May Come from Rational Beings, *New York Times,* April 13, 1965, pp. 1, 29.

Tesla, Nikola: Talking with the Planets, *Curr. Lit.,* p. 359, March 1901.

That Prospective Communication With Another Planet, *Curr. Opinion,* vol. 66, pp. 170–171, March 1919.

Thomas, L.: CETI, *New England J. Med.,* vol. 286, pp. 306–307, 1972.

A $3-Billion Radio Telescope Urged for Exploration of Space, *New York Times,* March 23, 1966, p. 33.

Todd, D.: Radio Messages from Mars, *Lit. Digest,* vol. 82, no. 10, p. 28, 1924.

Tovmasyan, G. M.: A Ring Telescope for Communication with Extraterrestrial Civilizations, in G. M. Tovmasyan (ed.), "Extraterrestrial Civilizations," NASA Scientific and Technical Information Facility translation, N67-30337, pp. 60–61.

Trench, Walter: correspondence in *Analog,* vol. 82, no. 4, pp. 168–169, 1968.

Troitskii, V. S.: Some Considerations on the Search for Intelligent Signals from Space, in G. M. Tovmasyan (ed.), "Extraterrestrial Civilizations," NASA Scientific and Technical Information Facility translation, N67-30338, pp. 62–71.

———: The Search for Signals of Extraterrestrial Life, Army Foreign Science and Technology Center, Washington, D.C., FSTC-HT-23-009-70; translation of *Aviatsiya i Kosmonavtika* (USSR), no. 8, pp. 77–80, 1968.

Tsvetikov, Alexis N.: 'Next Question' and K. E. Tsiolkovsky, correspondence concerning Tsiolkovsky's opinions about extraterrestrial societies, *Science,* vol. 131, no. 3403, pp. 872–873, 1960.

Universal Decoding Plan for Interstellar Messages, *Sci. News Lett.,* vol. 78, p. 265, Oct. 22, 1960.

Walker, J. C. G.: The Search for Signals from Extraterrestrial Civilizations, *Nature,* vol. 241, pp. 379–381, 1973.

Webb, J. A.: Detection of Intelligent Signals from Space, in A. G. W. Cameron (ed.), "Interstellar Communication," pp. 178–191, W. A. Benjamin, Inc., New York, 1963; *7th National Communications Symposium Record,* pp. 10–15, IRE Professional Group on Communications Systems, 1961.

Who's Calling, Please?, *Newsweek,* vol. 71, p. 114, Apr. 15, 1968.

Wiley, John P., Jr.: Is Anybody Out There?, *Natural History,* vol. 80, pp. 42–43, Dec. 1971.

———: Postscript to the Jupiter Mission, 'Sky Reporter' section of *Natural History,* vol. 81, no. 4, pp. 44–45, 1972.

Wilhelm, John and Fred Golden: Is There Life on Mars—or Beyond?, *Time,* vol. 98, no. 24, pp. 50–58, 1971.

Willard, Dennis: correspondence on pulsars as intelligent beacons, *Analog,* vol. 84, no. 1, p. 168, 1969.

Wolfert, I.: They Listen to the Language of the Universe, *Reader's Digest,* vol. 94, pp. 95–99, Feb. 1969.

B.   *Interstellar Communication Languages*

And a Message from Earth . . . , *Time,* vol. 101, p. 62, April 9, 1973.

Bracewell, Ronald N.: The Opening Message from an Extraterrestrial Probe, *Astronautics and Aeronautics,* vol. 11, no. 5, pp. 58–60, 1973.

Calvin, Melvin: Talking to Life on Other Worlds, *Sci. Digest,* vol. 53, pp. 19, 88–89, Jan. 1963.

Campbell, John W.: Hydrogen Isn't Cultural, in Harry Harrison (ed.), "Collected Editorials from *Analog,*" pp. 173–178, Doubleday and Company, Inc., Garden City, N.Y., 1966.

Freudenthal, Hans: "LINCOS: Design of a Language for Cosmic Intercourse," North-Holland Publishing Co., Amsterdam, 1960.

Gardner, Martin: The Ozma Problem, and Is the Ozma Problem Solved?, "The Ambidextrous Universe," chaps. 18 and 25, pp. 158–165, 237–243, New American Library, Inc., New York, 1969.

———: Extraterrestrial Communication, "Martin Gardner's Sixth Book of Mathematical Games from Scientific American," pp. 253–262, W. H. Freeman and Company, San Francisco, 1971.

Gilmore, C. P.: Decoding Messages from Outer Space, *Popular Science,* vol. 194, pp. 72–77 + , June 1969.

Gladkii, A. V.: Languages for Communication Between Different Civilizations (synopsis), in G. M. Tovmasyan (ed.), "Extraterrestrial Civilizations," pp. 95–96, NASA Scientific and Technical Information Facility translation.

Golomb, Solomon W.: Extraterrestrial Linguistics, *Astronautics,* vol. 6, no. 5, pp. 46–47, 95, 1961.

Hogben, Lancelot: Astraglossa or First Steps in Celestial Syntax, *J. Brit. Interplanetary Soc.,* vol. 11, no. 6, pp. 258–274, 1952.

———: Cosmical Language, (a review of Hans Freudenthal's "LINCOS: Design of a Language for Cosmic Intercourse"), *Nature,* vol. 192, no. 4805, pp. 826–827, 1961.

Kahn, David: Messages from Outer Space, "The Codebreakers," pp. 938–965, Macmillan Company, New York, 1967.

Lawton, A. T.: The Interpretation of Signals from Space, *Spaceflight,* vol. 15, no. 4, pp. 132–137, 1973.

Lilly, John C.: "Man and Dolphin," Doubleday and Company, Inc., Garden City, N.Y., 1961.

———: Interspecies Communication, "Yearbook of Science and Technology," pp. 279–281, McGraw-Hill Publishing Company, Inc., New York, 1962.

———: Communication with Extraterrestrial Intelligence, *IEEE Spectrum,* vol. 3, no. 3, pp. 159–160, 1965.

———: "Problems in Inter-Species Communication," talk given at AAS Symposium on Extraterrestrial Life, May 23–25, 1966, unpublished.

———: "The Mind of the Dolphin," Doubleday and Company, Inc., Garden City, N. Y., 1967.

Lunan, D.: Space Probe from Epsilon Boötis, *Spaceflight,* vol. 15, no. 4, pp. 122–131, 1973.

———: Message from a Star . . . , *Time,* vol. 101, pp. 59, 62, April 9, 1973.

Marsden, B. G.: "Astronomical Telegrams," Transactions of the International Astronomical Union, vol. 14A, Reidel Publishing Company, Dordrecht-Holland, 1970.

Mercer, D. M. A.: Messages to the Stars, *UNESCO Courier,* vol. 19, pp. 4–7 + , Jan. 1966.

Nieman, H. W. and C. W. Nieman: What Shall We Say to Mars? *Sci. Amer.,* vol. 122, no. 12, pp. 298, 312, 1920.

Oakley, C. O.: Math—Our Link to Space People, *Sci. Digest,* vol. 49, pp. 7–13, June 1961.

Pioneer F's Interstellar Message to the Future, *Sci. News,* vol. 101, no. 9, p. 135, 1972.

Pioneer Plaque to Identify Spacecraft Origin, *Aviation Week and Space Technology,* vol. 96, no. 9, p. 13, 1972.

Pryor, Hubert: Can You Decipher This Interplanetary Message?, *Sci. Digest,* vol. 60, no. 2, pp. 37–38, 1966.

Rorschach in Space, *Time,* vol. 99, p. 60, June 15, 1972.

Sagan, Carl, Linda Salzman Sagan, and Frank Drake: A Message from Earth, *Science,* vol. 175, pp. 881–884, Feb. 25, 1972.

Smith, R. F. W.: "Linguistic Problems in Extraterrestrial Communication," presented at the 15th Annual Meeting of the Audio Engineering Society, Oct. 14–18, 1963.

———: Communication with Extraterrestrial Beings Called Improbable Unless Man Can Signal in Two Systems of Thought, *Sci. Fortnightly,* Oct. 30, 1963, p. 8.

Sukhotin, B. V.: Methods of Message Decoding, in S. A. Kaplan (ed.), "Extraterrestrial Civilizations," pp. 133–213, NASA TT F-631, trans. from Russian, 1971.

———: "Problems of Decoding," International Astronautical Federation, paper, 23rd Int. Astronautical Congress, Vienna, Oct. 8–15, 1972.

Sullivan, Walter: To Let Others Know We Are Here, *New York Times,* News of the Week in Review, Feb. 27, 1972.

Wiley, J. P., Jr.: Communication Plaque on Pioneer 10, *Natural History,* vol. 81, no. 4, pp. 44–45, 1972.

III. Philosophical, Psychological, and Sociological Questions of Interstellar Transport, Interstellar Communication, and Extrasolar Intelligence

Aronoff, Samuel: From Chemical to Biological to Social Evolution, in "Interstellar Communication: Scientific Perspectives," Cyril Ponnamperuma and A. G. W. Cameron (eds.), chap. 6, Houghton Mifflin Company, Boston, 1974.

Ascher, Robert and Marcia Ascher: Interstellar Communication and Human Evolution, in A. G. W. Cameron (ed.), "Interstellar Communication," pp. 306–315, W. A. Benjamin, Inc., New York, 1963; *Nature,* vol. 193, pp. 940–941, March 10, 1962.

Bova, Ben: Galactic Geopolitics, *Analog,* vol. 88, no. 5, pp. 51–62, 1972.

Breig, Joseph A. and L. C. McHugh: Other Worlds for Man, *America,* vol. 104, pp. 294–297, Nov. 26, 1960.

Brewster, D.: "More Worlds Than One: The Creed of the Philosopher, and the Hope of the Christian," John C. Hotten, London, 1870.

Brookings Institution: The Implications of a Discovery of Extraterrestrial Life, "Proposed Studies on the Implications of Peaceful Space Activities for Human Affairs," prepared for NASA, Report of the Committee on Science and Astronautics, U.S. House of Representatives, March 24, 1961, 87th Congress, U.S. Government Printing Office, Washington, D.C., 1961.

Bruns, J. E.: Cosmolatry, *Catholic World,* vol. 191, pp. 283–287, Aug. 1960.

Carey, G. C. S.: Lewis and Space: Biology, Ecology, and Theology, *Commonweal,* vol. 69, pp. 100–101, Oct. 24, 1958.

Carr, A.: Take Me to Your Leader, *Homiletic and Pastoral Review* (New York), vol. 65, pp. 255–256, Dec. 1964.

Chamberlin, R. V.: Life in Other Worlds: A Study in the History of Opinion, *Bull. University of Utah,* vol. 22, no. 3, Feb. 1932.

Clarke, Arthur C.: The Challenge of the Spaceship, *J. Brit. Interplanetary Soc.,* vol. 6, no. 3, pp. 66–81, 1946.

———: On the Morality of Space, *Sat. Rev.,* vol. 40, pp. 8–10, 35–36, Oct. 5, 1957.

Cole, Dandridge M.: Macro-Life: Parts I and II, *Space World,* vol. 1, no. 10, pp. 16–19, 44, 46–48, 1961; vol. 10, no. 11, pp. 16–17, 48–58, 60, 1961.

———: "The Next Fifty Years in Space — Man and Maturity," General Electric Co. Internal Document, Missile and Space Division, Valley Forge, Pa., PIBD-30-4, 1963; "Beyond Tomorrow: The Next 50 Years in Space," Amherst Press, Amherst, Wisconsin, 1965.

Conway, J.: If There Is Life on Other Planets Wouldn't There Have to Be Polygenesis?, *Catholic Messenger,* vol. 82, p. 10, Aug. 6, 1964.

Cornford, F. M.: Innumerable Worlds in Pre-Socratic Philosophy, *Classical Quart.* (London), vol. 28, pp. 1–16, Jan. 1934.

Cottrell, John: The Great Moon Hoax, *Sci. Digest,* vol. 66, no. 1, pp. 40–44, 1969.

Coupe, C.: Are the Planets Inhabited?, *American Catholic Quart.,* vol. 31, no. 124, pp. 699–720, 1906.

Davis, C.: The Place of Christ, *Clergy Review* (London), vol. 45, pp. 706–718, Dec. 1960.

Dethier, V. G.: Life On Other Planets, *Catholic World,* vol. 198, pp. 245–250, Jan. 1964.

Dole, Stephen H.: "The Search for a Rationale for Interstellar Communications," paper for presentation at AAAS 132nd Annual Meeting, Berkeley, Calif., symposium, Dec. 30, 1965.

Drake, Frank D.: Great Goal of Space Science-Life Itself, *Astronautics and Aeronautics,* vol. 3, pp. 16–17, Jan. 1965.

Dugan, George: Priest Suggests Rational Beings Could Well Exist in Outer Space, *New York Times,* Aug. 7, 1960, p. 14.

Easton, W. B.: Space Travel and Space Theology, *Theology Today,* vol. 17, pp. 428–429, Jan. 1961.

Enzmann, R. D. and R. Miller: The Grand Design, in "Space Mission Planning in Planetary Environments," *Annals N.Y. Acad. Sci.,* vol. 140, art. 1, pp. 592–593, 1966.

Faith and Outer Space, *Time,* vol. 71, no. 13, p. 37, March 31, 1958.

Haas, Ward J.: The Biological Significance of the Space Effort, *Annals N.Y. Acad. Sci.,* vol. 140, art. 1, pp. 659–666, 1966.

Haley, Andrew G.: "Metalaw — Reassessment in the Light of Certain Views Expressed by the Chief Justice of the United States," AIAA preprint no. 63-279, AIAA Summer Meeting, Los Angeles, Calif., June 17–20, 1963.

Hinton, H. E.: Correspondence in *Analog* in response to Ben Bova's Galactic Geopolitics, vol. 89, no. 4, p. 171, 1972.

Hoyle, Fred: "Of Men and Galaxies," University of Washington Press, Seattle, 1964.

Joseph E. Seagram & Sons, Inc.: "Life in Other Worlds: A Symposium," on the 10th Anniversary of the Samuel Bronfman Foundation, including J. R. Killian, Jr., G. B. Kistiakowsky, D. N. Michael, Harlow Shapley, Otto Struve, Arnold Toynbee, Walter Cronkite, Chet Huntley, and William L. Laurence, 1961.

Khovanov, G. M.: Rates of Development of Civilizations and Their Forecasting, in S. A. Kaplan (ed.), "Extraterrestrial Civilizations," pp. 213–238, NASA TT F-631, trans. from Russian, 1971.

Kleinz, J. P.: Theology of Outer Space, *Columbia,* vol. 40, pp. 27–28, Oct. 1960.

Lewis, C. S.: Onward Christian Spacemen, *Catholic Digest,* vol. 27, pp. 90–95, Aug. 1963.

Life On Other Planets Is Called Compatible with Jewish Ideas, *New York Times,* April 10, 1966, p. 36.

Lynch, J. J.: Christians on Other Planets?, *Friar,* vol. 19, pp. 26–29, Jan. 1963.

———: "What Kind of Beings Would We Expect Our Extraterrestrial Neighbors to Be?, AAS Preprint 66-82, 1966.

Mankind Is Warned to Prepare for Discovery of Life in Space, *New York Times,* Dec. 15, 1960, p. 16.

McHugh, L.: Others Out Yonder, *America,* vol. 104, pp. 295–297, Nov. 26, 1960.

———: Life In Outer Space, *Sign,* vol. 41, pp. 26–29, Dec. 1961.

Meeting Extraterrestrials, *Sci. News Lett.,* vol. 70, no. 24, p. 382, 1956.

Messages from Space, *America,* vol. 111, no. 24, pp. 770–771, 1964.

Metalaw, *New Yorker,* vol. 32, p. 19, Dec. 29, 1956.

Michaud, Michael A. G.: Interstellar Negotiation, *Foreign Service J.,* vol. 49, no. 12, pp. 10–14, 29–30, 1972.

Miller, E. C.: Ethics and Space Travel, correspondence in *Spaceflight,* vol. 4, p. 139, July 1962.

Missionaries to Space, *Newsweek,* vol. 55, Feb. 15, 1960, p. 90.

Moore, Aidan: New Frontiers for Man, correspondence in *Spaceflight,* vol. 13, no. 12, pp. 478–479, 1971.

No Room for Christian Faith, *Sign,* vol. 36, p. 14, Nov. 1956.

Other-Worldly Faith, *Newsweek,* vol. 51, p. 64, March 24, 1958.

Other Worlds, Other Beings?, *Newsweek,* vol. 60, no. 15, pp. 112–115, Oct. 8, 1962.

Perego, A.: Rational Life Beyond the Earth, *Theology Digest,* vol. 7, pp. 177–178, Fall 1959.

Philosophical Speculations Concerning Life on Earth and in Outer Space, trans. from Russian, *Priroda,* no. 11, pp. 88–101, 1965, Joint Publications Research Service, Washington, D. C., JPRS-34259, TT-66-30700.

Pinnock, T. W.: Correspondence: "The Challenge of the Spaceship" by Arthur C. Clarke, *J. Brit. Interplanetary Soc.,* vol. 6, no. 4, p. 126, 1947.

Pittenger, Norman W.: Christianity and the Man on Mars, *Christian Century,* vol. 73, pp. 747–748, June 20, 1956.

Pohl, Frederik: The Uses of Star Flight, *Annals N. Y. Acad. Sci.,* vol. 140, pp. 667–669, 1966.

Puccetti, Roland: "Persons: The Possibility of Intelligent Life in the Universe," Macmillan and Co., Ltd., London, 1968.

Raible, Daniel C.: Rational Life in Outer Space?, *America,* vol. 103, pp. 532–535, Aug. 13, 1960.

———: Men from Other Planets?, *Catholic Digest,* vol. 25, pp. 104–108, Dec. 1960.

Robinson, G. S.: Ecological Foundations of Haley's Metalaw, *J. Brit. Interplanetary Soc.,* vol. 22, no. 4, pp. 266–274, 1969.

Ross, H. E.: A Contribution to Astrosociology, *Spaceflight,* vol. 6, no. 4, pp. 120–124, 1964.

Schwartz, Mortimer: Ethics and Extraterrestrial Life, correspondence in *Spaceflight,* vol. 5, no. 1, p. 36, 1963.

Searle, G. M.: Is There a Companion World to Our Own?, *Catholic World,* vol. 55, pp. 860–878, Sept. 1892.

Space Law (discussion of a paper by A. G. Haley), *J. Brit. Interplanetary Soc.,* vol. 16, no. 1, pp. 40–43, 1957.

Space Theology, *Time,* vol. 66, no. 12, p. 81, Sept. 19, 1955.

Stapledon, Olaf: Interplanetary Man, *J. Brit. Interplanetary Soc.,* vol. 7, no. 6, pp. 213–233, 1948.

Stetson, Harlan True: Has Life Any Cosmic Significance?, "Man and the Stars," pp. 191–200, McGraw-Hill Book Company, Inc., New York, 1930.

Sullivan, Walter: Biologist Says: "Without Our Life, the Universe Might Be, but Not Be Known," *New York Times,* March 8, 1966, p. 19.

The Type of Mind That Believes in Life on Other Worlds, *Curr. Opinion,* vol. 71, pp. 630–631, Nov. 1921.

Verplaetse, Juliaan: Magnified Analogy of Terrestrial Law in Outer Space?, *Zeitschrift für Luftrecht und Weltraumrechtsfragen,* vol. 19, pp. 140–145 (in German), Jan. 1, 1970.

Wagner, Bernard M.: Sociological Aspects of Exobiology, "Exobiology," *AAS Science and Technology Series,* vol. 19, pp. 117–130, 1967.

Williamson, A. A.: Speculation on the Cosmic Function of Life, *J. Wash. Acad. Sci.,* vol. 43, no. 10, pp. 305–311, 1953.

World Poll Finds Wide Belief in Life on Other Planets, *New York Times,* June 13, 1971, sect. 1.

Younghusband, Sir Francis: "Life in the Stars," John Murray (Publishers), Ltd., London, 1927.

Zubek, T. J.: Theological Questions on Space Creatures, *Amer. Ecclesiastical Rev.,* vol. 145, pp. 393–399, Dec. 1961.

IV. Bibliographies, Compendia, and Books Covering Multiple Topics of Extrasolar Intelligence and Interstellar Communication

A. *Bibliographies*

Forward, R. L.: Bibliography of Interstellar Travel and Communication, in Robert D. Enzmann (ed.), "Use of Space Systems for Planetary Geology and Geophysics," *AAS Science and Technology Series,* vol. 17, pp. 307–325, 1967.

Mallove, Eugene F. and Robert L. Forward: "Bibliography of Interstellar Travel and Communication," Hughes Research Laboratories, Malibu, Calif.: Research Report 439, May 1971, and Research Report 460, Nov. 1972.

National Aeronautics and Space Administration: "Extraterrestrial Life: A Bibliography," Part I: Report Literature (1952–1964), Part II: Published Literature (1900–1964), NASA SP-7015, Sept. 1964 and Dec. 1965.

Sable, Martin H.: "UFO Guide: 1947–1967," Rainbow Press Company, Beverly Hills, Calif., 1967.

Shneour, Elie A. and Eric A. Otteson: "Extraterrestrial Life: An Anthology and Bibliography," supplement to C. S. Pittendrigh (ed.), "Biology and the Exploration of Mars," National Academy of Sciences, Washington, D.C., 1965.

West, Martha W. and Cyril Ponnamperuma: Chemical Evolution and the Origin of Life: A Comprehensive Bibliography, *Space Life Sciences*, vol. 2, no. 2, pp. 225–295, 1970, and *Space Life Sciences*, vol. 3, no. 3, 293–304, 1972.

B. *Compendia*

Cameron, A. G. W. (ed.): "Interstellar Communication," W. A. Benjamin, Inc., New York, 1963.

Freundlich, Martin M. and Bernard M. Wagner (eds.): "Exobiology — The Search for Extraterrestrial Life," AAS and AAAS Symposium, New York, N. Y., Dec. 30, 1967, *American Astronautical Society, Proc.,* vol. 19, 1969.

Kaplan, S. A.: "Extra-Terrestrial Civilizations-2," Russian edition, Glavnaya Redaktsiya Fiziko-Matematicheskoi Literatury, Moscow, 1969; English edition, Keter Publishers, 1971.

———, N. S. Kardashev, B. N. Panovkin, L. M. Gindilis, B. V. Sukhotin, and G. M. Khovanov: "Extraterrestrial Civilizations: Problems of Interstellar Communications," (trans. from Russian, "Vnezemnye Tsivilizatsii: Problemy Mezhzvezdnoi Sviazi," Izdatel'stvo Nauka, Moscow, 1969), Israel Program for Scientific Translations, Ltd., Keter Press, Jerusalem, 1970; NASA TT F-631, trans. from Russian, 1971.

Krasovskiy, V. I. (ed.): "Space—Collection of Articles," 1963; trans. from Russian *Kosmos* (Moscow), no. 1, pp. 1–96, Air Force Systems Command, Wright-Patterson AFB, Ohio, Foreign Technology Division, FTD-MT-64-239, AD-608184.

Tovmasyan, G. M. (ed.): "Extraterrestrial Civilizations," Proc. of the 1st All-Union Conference on Extraterrestrial Civilizations and Interstellar Communication (Byurakan, USSR, May 20–23, 1964), NASA Scientific and Tech. Information Facility translation, N67-30330 through N67-30342, NASA TT F-438, TT 67-51373.

———: "Extraterrestrial Civilizations-1," Russian edition, Akadamiya Nauk Armyanskoi, USSR, 1965; English edition, Keter Publishers, 1967.

C. *Multiple Topics Books*

Allen, Tom: "The Quest: A Report on Extraterrestrial Life," Chilton Books, Philadelphia, 1965.

Anderson, Poul: "Is There Life on Other Worlds?," Collier Books, The Macmillan Co., New York, 1968.

Berrill, N. J.: "Worlds Without End—A Reflection on Planets, Life, and Time," Macmillan Company, New York, 1964.

Binder, Otto O.: "Riddles of Astronomy," chaps. 10, 11, and 12, pp. 135–186, Basic Books (Publishers), Inc., New York, 1964.

Biraud, Francois and Jean-Claude Ribes: "Le Dossier des Civilisations Extra-Terrestres," Fayard, Paris, 1970.

Bova, Ben: "Planets, Life, and LGM," Addison-Wesley Publishing Company, Inc., Reading, Massachusetts, 1970.

Clarke, Arthur C.: "The Challenge of the Spaceship," Ballantine Books, Inc., New York, 1961.

Drake, F. D.: "Intelligent Life in Space," Macmillan Company, New York, 1962.

Ehrensvärd, Gösta: "Man On Another World," University of Chicago Press, Chicago, 1965.

Firsoff, V. A.: "Life Beyond the Earth – A Study in Exobiology," Basic Books (Publishers), Inc., New York, 1963.

Gatland, Kenneth W. and Derek D. Dempster: "The Inhabited Universe," Fawcett Publications, Inc., Greenwich, Conn., 1963.

Heuer, Kenneth: "Men of Other Planets," Viking Press, New York, 1964.

Kaplan, S. A. (ed.): "Extraterrestrial Civilization: Problems of Interstellar Contact," Science Publishing House, Moscow, 1969.

MacGowan, Roger A.: On the Possibilities of the Existence of Extraterrestrial Intelligence, in Frederick I. Ordway, III (ed.), "Space Science and Technology," vol. 4, pp. 39–110, Academic Press, Inc., New York, 1962.

——— and Frederick I. Ordway, III: "Intelligence in the Universe," Prentice-Hall, Inc., Englewood Cliffs, N. J., 1966.

Macvey, John W.: "Alone in the Universe?," The Macmillan Company, New York, 1963; "Journey to Alpha Centauri," The Macmillan Company, New York, 1965.

———: "How We Will Reach the Stars," Collier-Macmillan Ltd., London, 1969.

Mamikunian, G. and M. H. Briggs (eds.): "Current Aspects of Exobiology," Pergamon Press, New York, 1965.

Moore, Patrick and David A. Hardy: "Challenge of the Stars," Mitchell Beazley, 1972.

Perelman, R. G.: "Goals and Means in the Conquest of Space," NASA TT F-595, trans. from Russian, 1970.

Ponnamperuma, Cyril (ed.): "Exobiology," North-Holland Publishing Co., Amsterdam, 1972.

Sagan, Carl (ed.): "Communication with Extraterrestrial Intelligence," The M.I.T. Press, Cambridge, Mass., 1973.

———: "The Cosmic Connection: An Extraterrestrial Perspective," Doubleday and Company, Inc., Garden City, N. Y., 1973.

Shklovskii, I. S. and Carl Sagan: "Intelligent Life in the Universe," Holden-Day, Inc., Publisher, San Francisco, 1966.

Strong, James G.: "Flight to the Stars," Hart Publishing Company, New York, 1965.

Sullivan, Walter: "We Are Not Alone," McGraw-Hill Publishing Company, Inc., New York, 1964.

Young, Richard S.: "Extraterrestrial Biology," Holt, Rinehart and Winston, Inc., New York, 1966.

V. Miscellaneous

Abell, G. O.: The Nearest Stars, "Exploration of the Universe," appendix 12, Holt, Rinehart, and Winston, Inc., New York, 1969.

Afshar, H. K.: The Innovative Consequences of Space Technology and the Problems of the Developing Countries, *Institute of Geophysics, Publication,* no. 56, Teheran University Press, Teheran, 1971.

Agel, Jerome: "The Making of Kubrick's '2001'," The New American Library, Inc., New York, 1970.

Allen, C. W.: The Nearest Stars, "Astrophysical Quantities," chap. 12, sect. 13, University of London Press, Ltd., 1962.

Berrill, N. J.: "The Origin of Life, and You and the Universe, "You and the Universe," chaps. 15 and 20, Dodd, Mead, and Company, Inc., New York, 1958.

———: The Search for Life, in Oscar H. Rechtschaffen (ed.), "Reflections on Space," pp. 37–45, U. S. Air Force Academy, Colorado, Jan. 1964.

Bobrovnikoff, N. T.: Soviet Attitudes Concerning the Existence of Life in Space, in G. E. Wukelic (ed.), "Handbook of Soviet Space-Science Research," pp. 453–472, Gordon and Breach, Science Publishers, Inc., New York, 1968.

Briton Asserts Man May Steer Planets, *New York Times,* June 5, 1965, p. 12.

Ciardi, John: Is Anyone There? (humor), *Sat. Rev.,* p. 27, Nov. 20, 1971.

Clarke, Arthur C.: The Sentinel, "Expedition to Earth," Ballantine Books, Inc., New York, 1953.

———: The Songs of Distant Earth, "The Other Side of The Sky," The New American Library of World Literature, Inc., New York, 1959.

———: Trouble in Aquila, and Other Astronomical Brainstorms, in I. J. Good (ed.), "The Scientist Speculates," pp. 235–238, William Heinemann, Ltd., London, 1962.

———: "Profiles of the Future," Harper and Row, Publishers, Inc., New York, 1963.

———: "2001: A Space Odyssey," New American Library, Inc., New York, 1968.

———: "The Lost Worlds of '2001'," New American Library, Inc., New York, 1972.

Cleaver, A. V.: A review of "Goals and Means in the Conquest of Space," (book by R. G. Perelman, NASA TT F-595, 1970), *Spaceflight,* vol. 13, no. 12, pp. 473–474, 1971.

von Däniken, Erich: "Chariots of the Gods?," Bantam Books, Inc., New York, 1971.

———: "Gods from Outer Space," Bantam Books, Inc., New York, 1972; "Zurück zu den Sternen (Return to the Stars), in German, Econ-Verlag Gmbh., 1968.

Drake, F. D.: Intelligent Life in the Universe (review of book by I. S. Shklovskii and C. Sagan), *Amer. Scientist,* vol. 55, pp. 300A, 302A, 1967.

Dyson, F. J.: Correspondence: reply to James R. Newman's review of "Interstellar Communication," *Sci. Amer.,* vol. 210, no. 4, pp. 8–9, 1964.

Edwards, D. F. A.: The Last Frontier, *Spaceflight,* vol. 12, no. 9, pp. 374–377, 1970.

A Fanciful Preview to New Facts, *Life,* vol. 41, no. 13, pp. 40–41, Sept. 24, 1956.

Flammarion, Camille: "La Pluralité des Mondes Habitués: Étude ou L'on Expose les Conditions d'Habitabilité des Terres Célèstes Discutées au Point de vue de l'Astronomie, de la Physiologie et de la Philosophie Naturelle," 28th edition, in French, Didier, Paris, 1880.

Fontenelle, B. LeBovier de (1657–1757): "Entretiens sur la Pluralité des Mondes. Digression sur les Anciens et les Modernes," Robert Shackleton (ed.), in French, Clarendon Press, Oxford, 1955.

Forward, Robert L.: Pluto—the Gateway to the Stars, *Missiles and Rockets,* vol. 10, pp. 26–28, April 2, 1962; *Sci. Digest,* vol. 52, pp. 70–75, Aug. 1962.

Froman, Darol: The Earth As a Man-Controlled Space Ship, *Phys. Today,* pp. 19–23, July 1962.

Gardner, Martin: "The Ambidextrous Universe," New American Library, Inc., New York, 1969.

Gatland, K. W.: New Frontiers, *Spaceflight,* vol. 12, no. 4, p. 166, 1970.

General Dynamics Corporation, Astronautical Division: "2063 A.D.: Prophecies by Distinguished Americans of Man's Employment of Space," 1963.

Gould, Jack: Review of television show, "We Are Not Alone," *New York Times,* Oct. 22, 1966, p. 63.

Hartlaub, G. F.: "Bewusstsein auf Anderen Sternen? Ein Kleiner Leitfaden Durch die Menscheiträme von den Planetenbewohnern," in German, Ernst Reinhardt Verlag, Munich, 1951.

Heinlein, Robert and Harold Wooster: Xenobiology, correspondence in *Science,* vol. 134, pp. 223, 225, July 21, 1961.

Hennes, John P., L. Gail Despain, and Jack L. Archer: "Scientific Goals of Missions Beyond the Solar System," Proc. of AAS 17th Annual Meeting, preprint no. AAS-71-163, June 28–30, 1971; *Adv. in the Astronautical Sciences — The Outer Solar System,* vol. 29, part II, pp. 597–616, 1900.

Hénon, M.: Communications Interstellaires, book review, *L'Astronomie* (Paris), vol. 78, no. 5, pp. 169–180, 1964.

Hermann, J.: "Leben auf Anderen Sternen?" (Life On Other Stars?), in German, C. Bertelsmann, Guütersloh, 1963.

Hoffman, Banesh: Ads From Space?, *New York Times,* April 14, 1968, p. 17.

Holmes, D. C.: "The Search for Life on Other Worlds," Sterling Publishing Co., Inc., New York, 1966.

Hoyle, Fred and John Elliott: "A For Andromeda," Harper and Brothers, New York, 1962.

Huang, Su-Shu: Life in Space and Humanity on the Earth. A Joint Review of Five Books, NASA TM X-56136, 1964.

Huygens, C.: "The Celestial Worlds Discovered; or Conjectures Concerning the Inhabitants, Plants, and Productions of the Worlds in the Planets," 2nd edition, James Knapton, Printer, at the Crown in St. Paul's Church-Yard, London, 1722.

van de Kamp, Peter: Stars Nearer than Five Parsecs, *Sky and Telescope,* vol. 14, no. 12, pp. 498–499, 1955.

———: Stars Nearer Than Five Parsecs, *Astronomical Soc. of the Pacific, Publications,* vol. 81, no. 478, pp. 5–10, 1969.

von Khuon, Ernst: "Waren Die Götter Astronauten?" (Were the Gods Astronauts?), in German, Econ-Verlag, Düsseldorf, 1970.

Kind, S. S.: Energy Fixation and Intelligent Life, *J. Brit. Interplanetary Soc.,* vol. 11, no. 4, pp. 168–172, 1952.

Kristofferson, Karl E.: First Voyage to the Stars, concerning Pioneer 10, *Reader's Digest,* pp. 83–87, Sept. 1972.

Kuiper, Gerard P.: The Nearest Stars, *Astrophys. J.,* vol. 95, pp. 201–212, Jan.–March 1942.

Lawton, A. T.: Review of "Challenge of the Stars" (by Patrick Moore and David A. Hardy), *Spaceflight,* vol. 14, no. 10, p. 397, 1972.

Le Poer Trench, Brinsley: "The Sky People," Neville Spearman, London, 1960.

Lederberg, Joshua: Exobiology, correspondence in *Science,* vol. 142, no. 3596, p. 1126, 1963.

Lethbridge, Thomas Charles: "The Legend of the Sons of God: A Fantasy?," Routledge and Kegan Paul, Ltd., London, 1972.

Ley, Willy: The Stars in Their Courses, book review of "Alone in the Universe" (by John W. Macvey), *New York Times,* Dec. 8, 1963, sect. 7, p. 34.

———: Is Anybody Listening?, book review of "We Are Not Alone" (by Walter Sullivan), *New York Times,* Dec. 20, 1964, sect. 7, p. 10.

———: Things That Might Be, book review of "Intelligent Life in the Universe" (by I. S. Shklovskii and Carl Sagan), *New York Times,* Dec. 18, 1966, sect. 7, p. 23.

Life Beyond Our Planet, *New York Times,* Dec. 16, 1960, p. 32.

Lovell, Sir Bernard: Some Reflections on Ethics and the Cosmos, "The Exploration of Outer Space," pp. 68–82, Harper and Row, Publishers, Inc., New York, 1962.

Luyten, Willem J.: Astrofantasies and Contracts, correspondence in *Science,* vol. 145, no. 3629, p. 23, 1964.

Man Not Alone in Cosmos, *Sci. News Lett.,* vol. 70, no. 11, p. 163, 1956.

Moore, Richard: "One Hundred Nearby *Sol*-Type Stars from the 'Yale Catalog of Parallaxes'," obtainable from Richard Moore, 69 Oak Hill View, Rochester 11, New York.

Mumford, G. S.: Is There Life on Earth?, *Sky and Telescope,* vol. 31, pp. 213–214, April 1966.

Murchland, Bernard: Review of "Persons" (by Roland Puccetti), *Commonweal,* vol. 91, no. 21, pp. 592–594, 1970.

Mutschall, Vladimir E.: "Soviet Long-Range Space-Exploration Program (Analytical Survey)," Aerospace Technology Div., Library of Congress Report (ATD 66–57), May 13, 1966.

Newman, James R.: Book review of "Interstellar Communication," A. G. W. Cameron (ed.), *Sci. Amer.,* vol. 210, pp. 141–146, Feb. 1964.

O'Brien, Katharine: Hello Out There (Project Ozma), poem in *Science,* vol. 131, no. 3408, p. 1263, 1960.

Öpik, E. J.: Life and Intelligence in the Universe: Bottomless Speculations, *Irish Astronomical J.,* vol. 8, pp. 128–139, 1967.

Ordway, Frederick I., III: "Life in Other Solar Systems," E. P. Dutton and Co., New York, 1965.

———: Some Implications of Extrasolar Intelligence, *Annals N. Y. Acad. Sci.,* vol. 140, art. 1, pp. 653–658, 1966.

Ovenden, Michael W.: "Life in the Universe: A Scientific Discussion," Doubleday and Company, Inc., Garden City, N. Y., 1962.

Page, Thornton and Lou Williams Page: Life in the Universe, "The Origin of the Solar System," p. 176, Macmillan Co., New York, 1966.

Prehoda, Robert W.: "Designing the Future—The Role of Technological Forecasting," Chilton Book Company, Philadelphia, 1967.

Prize Offered for Talk With Another Planet, *New York Times,* Oct. 9, 1963, p. 11.

Proell, Wayne: A New Map of Solar Space—Part I, *J. Spaceflight,* vol. 5, no. 8, 1953.

Pryor, Hubert: The Search for Extraterrestrial Life, *Sci. Digest,* vol. 60, no. 2, pp. 28–36, 39–41, 1966.

Prytz, John M.: Applied Exobiology, *Space World,* vol. H-11-95, pp. 34–35, Nov. 1971.

Review of "Bibliography of Interstellar Travel and Communication" (by Eugene F. Mallove and Robert L. Forward), *Spaceflight,* vol. 14, no. 6, p. 230, 1972.

del Rey, Lester, William Tenn, Poul Anderson, and A. E. van Vogt: Eight Eyes on Strange New Worlds, *Esquire,* pp. 56–59, Jan. 1966.

Ritner, Peter: "The Society of Space," chaps. 7 and 8, Macmillan Company, New York, 1961.

Rynin, N. A.: *Interplanetary Flight and Communication,* vol. 1, no. 2, *Spacecraft in Science Fiction* (fictional account of interplanetary flight and communication), Israel Program for Scientific Translations, Ltd., Jerusalem. Avail. NTIS, sponsored by NASA and NSF. Translation into English of *Mezhplanetnye Soobshcheniya,* vol. 1, no. 2, *Kosmicheskie Korabli,* Leningrad, 1928.

Sagan, Carl and David Wallace: A Search for Life on Earth at 100 Meter Resolution, *Icarus,* vol. 15, no. 3, pp. 515–554, 1971.

Sandage, Allan: The Stars Within 15 Parsecs of the Sun, in D. J. K. O'Connell (ed.), "Stellar Populations," pp. 287–302, Interscience Publishers, Inc., New York, 1958.

Sendy, Jean: "Those Gods Who Made Heaven and Earth," translated by Lowell Bair, Berkley Publishing Corporation, New York, 1972.

Seybold, P. G.: "A Survey of Exobiology," Rand Corporation RM-3178-PR, Santa Monica, Calif., 1963 (N63-13966).

Shapley, Harlow: "Of Stars and Men," Beacon Press, Boston, 1958.

———: "The View from a Distant Star," Dell Publishing Co., Inc., New York, 1967.

Soviet Is Told Planets Have Beings Who Think, *New York Times,* Feb. 21, 1960, p. 7.

Swedenborg, E.: "The Earths in Our Solar System, Which Are Called Planets, and the Earths in the Starry Heavens, Their Inhabitants and Angels Thence, from Things Heard and Seen" (originally in Latin, published in London, 1758), B. A. Whittenmore, Boston, 1928.